Industrial Ecology and Environmental Management

Volume 1

Series Editor

Jingzheng Ren, Department of Industrial and Systems Engineering, Hong Kong Polytechnic University, Hung Hom, Kowloon, Hong Kong

Industrial Ecology and Environmental Management have been recognized as two powerful disciplines for solving the global challenges such as the depletion of non-renewable resources, global warming and energy/water shortage.

This book series includes informative books with comprehensive texts focusing on both theory and practice of industrial ecology and environmental management, it aims to present the multi-disciplinary frontiers of methods, models and technologies in industrial ecology and environmental management for scientists, engineers and researcher from academia, industry, government and institutions. It aims to build a platform for the worldwide scientists, engineers and researcher to share the frontiers of the methods, models and technologies in industrial ecology and environmental management for achieving a more sustainable future.

The books series will include both the theoretical tools and the practical applications of industrial ecology and environmental management for sustainable energy, industry and society.

More information about this series at https://link.springer.com/bookseries/16615

Jingzheng Ren
Editor

Multi-Criteria Decision Analysis for Risk Assessment and Management

Springer

Editor
Jingzheng Ren
Department of Industrial and Systems Engineering
The Hong Kong Polytechnic University
Hong Kong, China

ISSN 2730-6070 ISSN 2730-6089 (electronic)
Industrial Ecology and Environmental Management
ISBN 978-3-030-78151-4 ISBN 978-3-030-78152-1 (eBook)
https://doi.org/10.1007/978-3-030-78152-1

This Springer imprint is published by the registered company Springer Nature Switzerland AG
The registered company address is: Gewerbestrasse 11, 6330 Cham, Switzerland

Contents

Contributors

Siti Nazifah Zainol Abidin Faculty of Computer and Mathematical Sciences, Universiti Teknologi Mara, Jasin, Melaka, Malaysia

Alin Emanuel Artene Faculty of Management in Production and Transportation, Politehnica University of Timisoara, Timisoara, Romania

Ahmad Syafadhli Abu Bakar Mathematics Division, Centre for Foundation Studies in Science, University of Malaya, Kuala Lumpur, Malaysia

Selcuk Cebi Industrial Engineering Department, Yildiz Technical University, Istanbul, Turkey

Erkan Celik Department of Transportation and Logistics, Istanbul University, Istanbul, Turkey

Ulas Cinar Occupational Health and Safety Education Application and Research Center, Canakkale Onsekiz Mart University, Canakkale, Turkey

Xianlong Ge School of Economics and Management, Chongqing Jiaotong University, Chongqing, China

Alexander Gegov School of Computing, University of Portsmouth, Portsmouth, UK

Muhammet Gul Department of Emergency Aid and Disaster Management, Munzur University, Tunceli, Turkey

Larisa Ivascu Faculty of Management in Production and Transportation, Politehnica University of Timisoara, Timisoara, Romania

Yuanzhi Jin School of Economics and Management, Chongqing Jiaotong University, Chongqing, China;
Department of Computer Technology and Information Engineering, Sanmenxia Polytechnic, Sanmenxia, China

Mohd Zahurin Mohamed Kamali Mathematics Division, Centre for Foundation Studies in Science, University of Malaya, Kuala Lumpur, Malaysia

Ku Muhammad Naim Ku Khalif Centre for Mathematical Sciences, College of Computing and Applied Sciences, Universiti Malaysia Pahang, Pahang, Malaysia

James J. H. Liou Department of Industrial Engineering and Management, National Taipei University of Technology, Taipei, Taiwan

Huai-Wei Lo Department of Business Administration, Chaoyang University of Technology, Taichung, Taiwan

Linmao Ma School of Management, South-Central University for Nationalities, Wuhan, China

David Fernando Muñoz Department of Industrial and Operations Engineering, Instituto Tecnológico Autónomo de México, Mexico City, Mexico

Jingzheng Ren Department of Industrial and Systems Engineering, The Hong Kong Polytechnic University, Hong Kong, China

Omer Faruk Ugurlu Mining Engineering Department, Istanbul University-Cerrahpasa, Istanbul, Turkey

Huijuan Xiao Department of Industrial and Systems Engineering, The Hong Kong Polytechnic University, Hong Kong, China

Melih Yucesan Department of Mechanical Engineering, Munzur University, Tunceli, Turkey

Long Zhang Department of Industrial and Systems Engineering, The Hong Kong Polytechnic University, Hong Kong, China;
School of Business, Xinyang Normal University, Xinyang, China

Chapter 1
Risk Assessment: Indicators and Organizational Models

Larisa Ivascu and Alin Emanuel Artene

Abstract The risk is encountered in every activity, operation, process, system or decision-making project. Given the importance of this concept at the individual and organizational level, this approach emphasizes the characteristics of the concept. This chapter aims to make an inventory of the concept of risk, its importance for organizations. This chapter emphasizes the importance of risk assessment in the risk management process. Risk management is an important step in the risk management process. Based on this argument, a series of qualitative and quantitative methods are presented. At the end of the chapter, organizational methods and models are presented. The last part presents a selection of indicators that are used in the automotive industry.

Keywords Sustainability · Probability · Opportunity · Employee · Value · Risk · Risk management · Tool · Risk matrix

1.1 Introduction

Risk is present in any activity, process, or organization. From this perspective, special attention must be paid to risks. Organizational risks are risks that positively or negatively affect organizational activity. The risk is the chance or probability that a person will be harmed or have a negative effect on health if that risk occurs. The risk can lead to various organizational or personal losses (The Institute of Risk Management, 2002; Tang 2006; (IRM), 2010; Rass et al. 2020). The risk arises because of the existence of a source of danger or a hazard. These sources of danger contribute to the occurrence of risks. Risks can come from a variety of sources, including uncertainty in financial, legal, legislative data, errors in strategic planning, or other organizational decisions. The occurrence of risks develops a series of consequences that can

L. Ivascu (✉) · A. E. Artene
Faculty of Management in Production and Transportation, Politehnica University of Timisoara, Timisoara, Romania
e-mail: larisa.ivascu@upt.ro

A. E. Artene
e-mail: alin.artene@upt.ro

J. Ren (ed.), *Multi-Criteria Decision Analysis for Risk Assessment and Management*, Industrial Ecology and Environmental Management 1,
https://doi.org/10.1007/978-3-030-78152-1_1

affect the organizational activity. To reduce the negative impact of organizational risks and increase the risk of opportunity, risk management has an important role in the organizational environment. Risk management is the process by which risks are identified, assessed, treated and controlled in order to reduce the negative impact (Van Der Walt 2003; Luko 2013; Eaton 2015).

At the level of organizations, in the current conditions of the economy, the security and safety of information technology, human resources risks, all risks related to organizational data, risk management strategies have become the priorities of digitized organizations (Latham and Braun 2009; Arena et al. 2010; Teece et al. 2016). A strategic plan for risk management includes an early identification of sources of danger and the application of measures to eliminate them. For the consequences, methods are applied to reduce the impact, so that the organizational losses are reduced as much as possible.

At the organizational level, there is also the risk generated by unexpected events that can contribute to high costs for the organization or even the closure of the activity. Effective risk management can anticipate certain risks and harmful situations so that the losses recorded by the organization are as low as possible (Dahl 2011; Lalonde and Boiral 2012).

Therefore, the risk manager is very important at the level of any organization, regardless of the field of activity. From this perspective, the present chapter aims to present a definition of risk, to identify the entities involved in the process of organizational risks, highlighting the opportunity risk, the importance of efficient risk management and a series of organizational models existing in present. At the end of the chapter is presented a framework approach for assessing technical and technological risks in the automotive industry. This approach includes several indicators. The methodology used to define the risk indicators used in the proposed framework is described by the following steps: (a) the systemic mapping study through the study of 10 technical reports in the automotive field; (b) definition of indicators; (c) the proposal of the chair for evaluation.

1.2 Risk Definition

Defining the notion of risk is a complex task, given the diversity of meanings regarding risk, as well as the rich typology of risk at the enterprise level. Therefore, the following is the definition of this concept from the perspectives of several authors in the literature (Latham and Braun 2009).

The explanatory dictionary of the Romanian language highlights the risk as the probability of suffering a damage. In the Concise Oxford English Dictionary, risk is defined as "hazard, the possibility of a negative consequence, loss or exposure to chance" (McCracken 2003; Oxford English Dictionary 2017).

Before 1995, all official publications in the field of risk management used the negative connotation of risk, the term being associated with danger, trouble, loss, etc. In these definitions, risk was associated with uncertainty, with negative effects/loss/degradation of the company's objectives, so the risk was equivalent to danger (Adler and Dumas 1984; Fischhoff et al. 1984; Mayer et al. 1995; Šotić and Rajić 2015).

After 1995, the definitions of risk were neutral, with risk being defined as "uncertainty that may affect one or more of the company's objectives" or "uncertainty that may positively or negatively affect one or more of the company's objectives." These definitions give a positive or negative note to the notion (Dionne 2013; Management and Guides 2014; Heckmann et al. 2015).

After 2000, the literature presents risk as the combination of opportunity and threat notion (Dionne 2013; Management and Guides 2014; Izvercian and Ivascu 2014; Heckmann et al. 2015; Izvercian and Ivascu 2014).

Risk is a probability, a mathematical quantity that can be measured, calculated, and estimated. Risk is not a bad concept, risks are essential in the progress of a system, and failures are key elements of learning. Thus, the negative consequences must be balanced with the potential benefits associated with this opportunity (Izvercian and Ivascu 2014).

The risk involves an imprecise situation, condition or multi-dimensional measure of exposure to imprecise losses (Fan and Stevenson 2018). Also define risk as "any problem that influences an organization's ability to meet its objectives." The authors also argue that directors and administrators talk about competitiveness risk, market risk, financial risk, operational risk, technological risk, environmental risk, regulatory risk, legal risk, reputational risk, and political risk (The Institute of Risk Management, 2002; Van Der Walt 2003; Izvercian and Ivascu 2014; Rass et al. 2020).

In Browning's view, risk is an undesirable situation or circumstance that involves the likelihood of potentially negative consequences for the system. Technical risk refers to the risk associated with the system that will fail to meet its criteria performance (Dionne 2013; Management and Guides 2014; Izvercian and Ivascu 2014; Heckmann et al. 2015).

Risk is based on a complex of factors and can be identified as an inherent component that can have interventions in any of the organizational activities. Risk assessment is a component of strategic management and is important in any activity, regardless of the field of activity because it can have interventions at every organizational level. Risk assessment involves the following steps: risk identification, risk analysis and assessment, determination of priority interventions for risk limitation and risk treatment (Izvercian and Ivascu 2014; Heckmann et al. 2015).

The risk implies uncertainty and the presence of risk is felt in the distribution of possible outputs of the system (The Institute of Risk Management, 2002; (IRM), 2010).

Another vision of risk is presented as "any event that affects the ability of the company to meet its objectives" (Luko 2013; Soin and Collier 2013). Therefore, there is a diversity of approaches to the concept of risk in the literature, so that an analysis of them becomes relevant in the present research. Based on the specialized literature, the following catch-up definition can be outlined:

Organizational risk can affect the achievement of organizational objectives, so implicitly the achievement of the mission and the achievement of the assumed vision. Risks are generated by hazards that exist within the organization, and they can generate consequences with negative impact or various opportunities (Izvercian and Ivascu 2014).

From these definitions the two connotations can be outlined: the positive connotation and the negative connotation. So when we talk about risk, the following concepts are emphasized:

(a) Positive connotation—risk can contribute to the emergence of positive aspects that favorably affect the organization and even increase the level of competitiveness.
(b) Negative connotation—the appearance of richness in organizations can contribute to negative aspects whose impact is dependent on strategic management at the organizational level.
(c) Probability—it refers to the probability of occurrence of risks in organizations.
(d) Opportunity—the appearance of organizational risks contributes to the appearance of some opportunities, positive aspect that must be approached correctly by the organization.
(e) Severity—the existence of a hazard or a source of danger within the organization develops a series of negative consequences. The severity of these consequences is directly proportional to the strategic measures for mitigating hazards at the organizational level. If the organization applies the most efficient strategies to mitigate or eliminate the sources of danger, then the severity of the consequences is diminished.
(f) Hazard or source of danger—it is the source that can cause damage of various types at the organizational level. This includes electricity, noise, flooring, chemicals, and others.
(g) Value—it is estimated the losses that the occurrence of a risk would generate at organizational level having as reference the normal conditions for carrying out the activity.

Through these connotations, risk can develop several opportunities. That is, the concept of opportunity risk can be outlined.

1.3 Risk Opportunity

Events caused by risks can have negative, positive or both consequences. Risks are associated with events that have a negative impact, and organizational opportunities are identified with positive events generated by the occurrence of a risk. Opportunities represent the possibility that an event that will take place will positively affect the achievement of objectives, in support of value creation or conservation (Vassilev and Dimitrova 2015; Słowikowski et al. 2017). Organizational opportunities positively affect the achievement of objectives, the achievement of the mission, and the overcoming of the organizational vision (Luko 2013).

Table 1.1 The differences between risk and opportunity risk

Opportunity risk	Risk
The possibility of the event to "affect" the company positively and not negatively	The impact of the risk on the company can be negative or positive
Creates added value for the enterprise	The results cannot be known exactly
The value of the results may be higher than expected	The value of the results may be lower than expected
Contributes to achieving organizational goals	It can affect the achievement of organizational goals
Fulfill the mission of the organization in order to achieve the vision	It can affect the fulfillment of the organizational mission and the achievement of the organizational vision
Requests immediate action from the organization to record positive results	The undertaking of activities depends on the risk management plan
It involves measures to strengthen the positive effects at the organizational level	It involves measures to eliminate hazards
It requires an integrated approach to opportunities with set goals to increase the impact of positive consequences	It involves measures to reduce the severity of the consequences
Uncertainty affecting risk opportunities may limit the benefits of the organizational approach	Uncertainty affecting data contributes to the increased impact of the consequences

There are several differences between risk and opportunity risk. This assessment is presented in Table 1.1.

1.4 International Standard ISO 31,000

International Organization for Standardization (ISO) is a non-governmental, international, and independent association that develops international standards designed to ensure the quality, safety, and efficiency of activities, operations or organizational processes. These developed standards are published, and organizations can implement them if they meet the criteria set out in the standard (ISO - The International Organization for Standardization 2009).

When we talk about risk management, we approach ISO 31,000, Risk management—Guidelines, provides principles, a framework, and a process for managing risk. This standard can be implemented by any organization, regardless of location, size or field of activity. This standard contributes to the achievement of organizational objectives and helps to identify opportunities and reduce the severity of the consequences. Provides directions for internal or external audit. If an organization wants to apply the directions of this standard, it can be ensured that it has an international reference point with which to compare its situation. This reference provides consistent principles for developing sound and consistent strategies. The whole approach is sustainable and can contribute to increasing organizational competitiveness (ISO - The International Organization for Standardization 2009). The benefits of adopting ISO 31,000 are presented in Table 1.2.

Table 1.2 Organizational benefits generated by ISO 31,000

ISO	Organizational benefits
ISO—Risk management—Guidelines, provides principles	An international reference for comparing the organizational situation
	A recognized reference for the organization
	Simple terminology for organizations in different fields of activity
	Contributes to continuous improvement at the organizational level
	It has strong links with governance structures and decisions
	Accentuates the improvement of the quality level
	Emphasizes the integration, design, implementation, evaluation and improvement of activities, processes and organizational systems
	It does not impose solid knowledge in the field of risk management
	It strikes a balance between the stages of risk management and the business imperative
	Provides a guidance framework suitable for any company
	The flexibility of the standard contributes to an easy implementation
	Providing an appropriate set of tools that apply to the entire organization is a major advantage
	It allows efficient communication at every level
	Allows quick updating of information
	Provides a dynamic view of organizational risks
	Allows users to view tartar and control pads in accordance with organizational objectives
	It offers the perfect integration of some methods used in risk management (bow-tie, check lists, sensitivity analysis, cost-benefits analysis and others)
	Contributes to the efficient application of organizational risk policies

1.5 Risk Management Process

Risk management is an organized process for identifying what may be wrong, quantifying and assessing the associated risks, implementing measures related to actions to prevent or treat each identified risk (Van Der Walt 2003; Wohl and Harvey 2005; Eaton 2015). Risk Management is the process that plans, identifies, evaluates, controls, and communicates aspects related to potential risks in order to minimize the negative impact it can have on the organization. These stages can be structured as follows:

(1) Risk Planning—it is the stage in which each organization plans its activity for the risk management process. Before risks can be identified, assessed, and addressed, a plan or strategy must be developed. Planning is part of the basic risk management process, including: (1) developing and documenting an organized, comprehensive, and interactive risk management strategy; (2) determining the methods to be used in the risk management strategy; (3) planning adequate resources. Risk planning is iterative and includes the scheduling of assessment, management, and monitoring activities and processes. The result of this action is called the risk management plan.

(2) Risk Identification—it is the stage in which the risks that could affect the achievement of organizational objectives, the fulfillment of the mission, and the achievement of the vision are determined. At the same time, a correct identification of risks can contribute to the realization of investments and of the activities and organizational processes. Within this stage a number of methods can be used, for example: Delphi method, brainstorming, documentation reviews, interviews, SWOT analysis (Strengths, Weakness, Opportunities and Threats), historical evaluation of the organization, and use of checklist analysis and others.

(3) Risk Assessment—it is in this stage are analyzed and evaluates the risks associated with a hazard or a source of danger. Qualitative risk analysis is the process of conducting a qualitative assessment of the risks identified within the enterprise. This stage sets a risk priority, depending on their potential effect on the company's objectives. Within this method, a series of methods can be used, for example, what-if analysis, fault tree analysis, failure mode event analysis, hazard operability analysis, Bow-Tie incident, event-tree, and others.

(4) Risk Controlling—it is the stage in which a series of risk control methods are applied. There are a number of possibilities for risk control, such as avoidance, loss prevention, loss reduction, separation, isolation, du-plication, or diversification. Associated with these control methods are the attitudes of managers that are corroborated with temperament, biological component, and attitude toward risks.

(5) Risk Communication—it is the stage in which the communication between the parties involved in risk management is carried out in order to increase the level of information and to better understand the risks and to apply the most efficient management methods.

Fig. 1.1 Stages of the risk management process

Therefore, risk management is a cyclical and continuous process that represents the coordination of activities to identify hazards, assess, control, monitor, and treat risks to achieve a balance between costs and benefits and achieve the company's objectives (Słowikowski et al. 2017). These stages are in the form of a cyclical process that takes place at the organizational level, Figure 1.1.

The following differences must be made in risk management.

Risk assessment—is the stage in which the nature of the risk is defined, the probability of occurrence, and the consequences that develop after its occurrence. This stage can be achieved by qualitative or quantitative methods.

Risk management—is the process by which a series of actions are taken to accept, assume and manage risk.

Risk Communication—is the stage in which a series of information exchanges is carried out to better understand the organizational risks.

1.6 Organizational Methods and Models for Risk Assessment

The purpose of risk assessment is to assess and describe the risks associated with an organizational decision-making problem. As presented in the previous chapter, risk assessment can be performed qualitatively and quantitatively. Qualitative methods use qualitative terms of appreciation, being a non-numerical method. Quantitative methods use numbers to express the level of risk. In continuation, Table 1.3, a selection of quantitative and qualitative methods that can be used is presented (Aloini et al. 2012; Theoharidou et al. 2012; Rausand 2013; Lefèvre et al. 2014; Abd Rashid

Table 1.3 Methods used for risk assessment

Method	Description	Advantage
Risk matrix approach	It is a qualitative method, applicable for various fields of activity	Consider the probability of occurrence and the severity of the consequences
Brainstorming	It is a method that involves debates with people with experience in risk management	It is a qualitative method that can lead to complex results in situations in which risk managers or experienced people take part in the debate
Indicator-based approach	The evaluation is based on indicators	These indicators are established according to the field of activity and contribute to a good evaluation
What-if analysis	The method for identifying hazards and hazardous situations, hazards and other elements that would lead to unintended consequences	It requires a deep understanding of the process and shows good results if the team involved has solid knowledge in this field
Fault tree analysis (FTA)	FTA is a deductive procedure that determines the different combinations of elements that can contribute to unwanted events	The graphic tree is a graphical method that logically presents a series of combinations that can lead to an unwanted event
Failure mode event analysis (FMEA)	It is a tree used in the evaluation of all possible defects	It is used in design to prevent defects that may occur
Event-tree analysis (ETA)	Inductive method that is applied from the bottom up	The success of applying this method depends on the amount of data held and the knowledge of the team
Monte Carlo	The application of the method involves the use of calculation algorithms	The results obtained are numerical and representative for risk managers
Scenario analysis	Different hypothetical evolutions are simulated	Simulations can develop effective solutions
Bow-tie method	It is a graphical method that evaluates hazards and the consequences of a risk	It is an effective method if the evaluation team has solid knowledge in this field
Efficiency of controls	Aims at the risks that may arise in a decision-making situation	Contributes to the treatment of a risk if the controls have been carried out effectively
Sensitivity analysis	Quantitative method that evaluates changes in a variable in a model	Contributes to the decision-making process through the result obtained after investigating the impact of changes in project variables

(continued)

Table 1.3 (continued)

Method	Description	Advantage
Cost benefit analysis	The method used by many companies to predict expected costs	Weigh the expected costs relative to the expected benefits
Checklist	It is the method used by many companies, being an accessible method	The evaluation team has an important contribution to the success of the evaluation. Historical data is used

and Yusoff 2015; Rivero et al. 2015; SheikAllavudeen and Sankar 2015; Chander and Cavatorta 2017).

1.7 Defining Indicators for Risk Assessment in the Manufacturing Industry

The Romanian manufacturing industry has an important share in the national economy. The number of employees is increasing, and the attractiveness of this field is high. The situation of the Romanian industries is presented in Table 1.4. For the calculation of the value indices, the year 2015 was used as a reference. The number of enterprises by fields of activity is presented in Appendix 1. These data were collected from the National Institute of Statistics, being public data, displayed in the statistical database. The manufacturing industry has an important share in the economic activity, presenting a series of innovations and a high number of jobs (Institutul Național de Statistică 2019).

From the perspective of the number of employees, the complete situation is presented in Appendix 2. For industry, the situation of employees is presented in Table 1.5. The manufacturing industry registers the most employees.

Table 1.4 Value indices of turnover in industry divisions

Industry divisions	January of each year evaluated						
	2015	2016	2017	2018	2019	2020	October 2020
	Percent	Percent	Percent	Percent	Percent	Percent	Percent
Total	84,2	79,3	85,7	101,4	108,2	118,4	143,7
Extractive industry	118,3	99,7	101,6	94,8	109,5	92,5	104,3
Manufacturing industry	81,7	77,8	84,6	101,9	108,1	120,3	146,6

Source NIS, 2020

Table 1.5 Number of employees in industry divisions

Industry divisions	Year					
	2015	2016	2017	2018	2019	2020
Extractive industry	80,762	57,863	54,022	51,447	49,025	48,313
Manufacturing industry	1,341,847	1,122,321	1,145,001	1,195,118	1,205,129	1,192,979

Source NIS, 2020

As a result, the manufacturing industry is an important one, being selected for the study of indicators for risk assessment. Below, Table 1.6, is the result of the stages of identification of the indicators based on the three stages carried out: (a) the systemic mapping study through the study of 10 technical reports in the automotive field; (b) definition of indicators; (c) the proposal of the chair for evaluation.

These are the general risk categories that have been identified in the reports of companies in the automotive industry. These are part of an exploratory study, being a reference for further developments.

1.8 Case Study

To highlight the case study, this section presents in the first part how to apply these indicators, and in the second part are evaluated a series of organizational models for risk assessment. The 20 indicators proposed in the next section are used for one of the large companies operating in the automotive industry. The evaluated company carries out its activity globally, being one of the World leaders in the field of car construction. The following severity classes and probability classes are used to evaluate the proposed indicators. These are presented in Tables 1.7 and 1.8.

The definition of risk levels is presented in Table 1.9. These levels are used in the risk matrix that is used for risk communication. The level of risk is calculated as the product between the severity class and the probability class.

The following is an example of the use of risk assessors. This activity is presented in Table 1.10. The severity classes and the probability classes defined above were used. Depending on the level of risk, the anticipated actions are initiated.

Organizations depending on the field of activity can select the application of a qualitative or quantitative method. Evaluating organizational models for risk assessment can highlight a few examples.[1]

[1] Online Platforme, RiskWatch, available online on https://riskwatch.com.

Platforma Capra, Probabilistic Risk Assessment, available online on https://ecapra.org.

Platforma OiRA, Online interactive Risk Assessment, available online on https://oiraproject.eu/en.

Table 1.6 Indicators for risk assessment in the automotive industry

No	Risk category	Hazard description
1	Air temperature	Use appropriate equipment
2	Defective performance of operations (commands, maneuvers, positioning, fixing, etc.)	Failure to follow the work instruction by performing operations that are not provided in the procedure
3	Design of bodies or particles	In case of accidental detonation of the gas generator Pallet trucks, forklifts, moving parts
4	Elbows out	When placing the parts in the support—shoulders raised
5	Failure to perform operations	Failure to perform operations when assembling parts Use of the electric forklift and/or hand pallet truck. Acting them and manipulating them
6	Falls from the same level (imbalance, slip, obstruction)	Obstruction, imbalance, slipping in the workplace area; during travel to and from the workplace (e.g., from the locker room, social groups, dining room, etc.)
7	Improper working method (non-observance of work instructions)	Failure to follow the work instruction by performing operations that are not provided in the procedure Parts, subassemblies, packaging
8	Indirect electric shock	Feeding/removing the finished products from/on the site, the specially arranged areas outside the section The production lines are supplied with electricity
9	Manual handling of masses	In case of accidental detonation of the gas generator
10	Movements under the effect of gravity (falling, sliding, rolling—objects)	Parts, subassemblies that may fall Pallet trucks, forklifts, moving parts Using the cutter when folding cardboard packaging
11	Omission of operations	Failure to perform operations when assembling parts
12	Personal protective equipment is used for the purpose for which it was granted	Use of the electric forklift and/or hand pallet truck. Acting them and manipulating them
13	Pressure vessels	The production lines are operated with compressed air—there are bottles with compressed air at each production line
14	Sharp objects/edges	Loading the electric forklift Sharp edges of parts and subassemblies
15	Sitting vs. standing	Orthostatic working position

(continued)

Table 1.6 (continued)

No	Risk category	Hazard description
16	Start/stop of technical equipment	Starting the machines for assembling the components to make the product
17	The work equipment is equipped with protection systems	Protective barriers
18	Travel/stationary in dangerous areas	The traffic on the access roads from the location, on which the supply of the lines is made. Travel inside the section
19	Glossy surfaces	Glossy surfaces can lead to operator slippage or production deficiencies
20	Large equipment	Large equipment (involving risky operations) must have operating procedures to avoid accidents with severe consequences

Table 1.7 Defining severity classes

Severity Class (SC)	Consequences	Description
1	Small	Event followed by temporary incapacity for work of maximum 2 days
2	Medium	Event followed by temporary incapacity for work between 3 and 44 days
3	Big	Event followed by temporary incapacity for work between 45 and 180 days
4	Very big	Event ended with disability
5	Maxim	Event ended with death

Table 1.8 Defining probability classes

Probability Class (PC)	Frequencies	Description
1	Extremely rare	The frequency of occurrence of an event is greater than 10 years
2	Rarely	The frequency of occurrence of an event is between 5 and 10 years
3	Rare	The frequency of occurrence of an event is between 1 year and 5 years
4	Frequent	The frequency of occurrence of an event is between 1 month and 1 year
5	Very frequently	The frequency of occurrence of an event is less than 1 month

Table 1.9 Defining risk levels

Risk level	Actions
1	No action is required
2–6	Medium and long-term corrective actions are required (1–5 years, over 5 years)
7–10	Short-term corrective actions are required (maximum 1 month)
11–16	Urgent corrective actions are required
Greater than 16	The activity stops

(a) OnRisk—it is an online platform for identifying technical and technological risks. The development of this risk assessment platform that includes innovative elements and reduced resources has the following main purposes: complete description of the evaluated object, detailed planning of the whole process, communication between the company's stakeholders, identification of hazards within the evaluated system, complete documentation of the evaluation process, interpretation of the obtained results, continuous monitoring and control (Ivascu et al. 2015; Gaureanu et al. 2016).

(b) RiskWatch—it is available online and offers solutions for different industries. It integrates several graphical features that can be easily interpreted (Fabbri 2020).

(c) Goat—online platform for risk assessment. The risk assessment is based on the probability of occurrence and targets a few areas of activity (Aven 2016).

(d) OiRA—interactive platform for occupational risk assessment. This platform covers several areas of activity and an evaluation is performed based on a checklist (Dudley et al. 2007; Василев 2014; Vassilev and Dimitrova 2015; Słowikowski et al. 2017).

1.9 Conclusions

Risk is present in all companies and must be addressed correctly to contribute to the achievement of organizational objectives. It has a positive connotation and a negative connotation. There are several methods, models, frameworks, and techniques that can be used in the risk management process. Their application depends on the management team and on certain particularities of the field of activity. A correct approach to organizational risks contributes to the identification of opportunities that can improve the organizational situation.

Table 1.10 Risk assessment for an automotive company

No	Risk category	Hazard description	SC	PC	Measure
1	Air temperature	Use appropriate equipment	4	5	Additional equipment
2	Defective performance of operations (commands, maneuvers, positioning, fixing, etc.)	Failure to follow the work instruction by performing operations that are not provided in the procedure	3	4	Reformulations and completions of the procedures
3	Design of bodies or particles	In case of accidental detonation of the gas generator Pallet trucks, forklifts, moving parts	3	3	Additional information
4	Elbows out	When placing the parts in the support —shoulders raised	3	4	Regulation of breaks in the work process
5	Failure to perform operations	Failure to perform operations when assembling parts Use of the electric forklift and/or hand pallet truck. Acting them and manipulating them	3	2	Supplementary information and training
6	Falls from the same level (imbalance, slip, obstruction)	Obstruction, imbalance, slipping in the workplace area; during travel to and from the workplace (e.g., from the locker room, social groups, dining room, etc.)	2	2	Supplementing safety features for employees
7	Improper working method (non-observance of work instructions)	Failure to follow the work instruction by performing operations that are not provided in the procedure Parts, subassemblies, packaging	2	2	Video supervision of employees
8	Indirect electric shock	Feeding/removing the finished products from/on the site, the specially arranged areas outside the section The production lines are supplied with electricity	3	3	Protection equipment
9	Manual handling of masses	In case of accidental detonation of the gas generator	2	2	Protection equipment

(continued)

Table 1.10 (continued)

No	Risk category	Hazard description	SC	PC	Measure
10	Movements under the effect of gravity (falling, sliding, rolling—objects)	Parts, subassemblies that may fall Pallet trucks, forklifts, moving parts Using the cutter when folding cardboard packaging	3	2	Protective equipment and video monitoring
11	Omission of operations	Failure to perform operations when assembling parts	4	4	Video monitoring
12	Personal protective equipment is used for the purpose for which it was granted	Use of the electric forklift and/or hand pallet truck. Acting them and manipulating them	3	3	Information
13	Pressure vessels	The production lines are operated with compressed air—there are bottles with compressed air at each production line	4	4	Performing preventive maintenance
14	Sharp objects/edges	Loading the electric forklift Sharp edges of parts and subassemblies	3	4	Equipped with protective gloves
15	Sitting vs. standing	Orthostatic working position	3	3	Regulating the alternation of activity within the technological process
16	Start/stop of technical equipment	Starting the machines for assembling the components to make the product	2	1	Corrective information
17	The work equipment is equipped with protection systems	Protective barriers	3	2	Performing preventive maintenance
18	Travel/stationary in dangerous areas	The traffic on the access roads from the location, on which the supply of the lines is made. Travel inside the section	2	3	Warning with graphic symbols
19	Glossy surfaces	Glossy surfaces can lead to operator slippage or production deficiencies	4	3	Corrective information
20	Large equipment	Large equipment (involving risky operations) must have operating procedures to avoid accidents with severe consequences	3	3	Performing preventive maintenance

Appendix A

See Tables A.1 and A.2.

Table A.1 Number of active enterprises by field of activity

Economic activities	Year					
	2015	2016	2017	2018	2019	2020
	Number	Number	Number	Number	Number	Number
Total	554,967	507,440	513,850	527,792	553,796	576,545
Agriculture, forestry and fishing	13,602	17,471	18,396	19,139	19,916	20,514
Extractive industry	1083	1112	1107	1076	1014	1033
Manufacturing industry	57,305	48,090	48,404	48,347	49,837	52,451
Production and supply of electricity and heat, gas, hot water and air conditioning	506	1503	1460	1350	1206	1200
Water distribution; sanitation, waste management, decontamination activities	2366	3160	3049	2968	3022	3074
Construction	59,389	47,814	48,341	49,716	52,792	55,978
Wholesale and retail trade; repair of motor vehicles and motorcycles	214,137	176,202	171,959	169,712	172,435	172,856
Transport and storage	34,489	39,666	41,746	44,504	48,382	51,944
Hotels and restaurants	23,653	25,111	25,497	25,612	26,414	27,182
Information and communications	20,049	19,499	20,619	22,012	23,837	25,452
Financial intermediation and insurance	6840	6903	7244	8225	8220	8461
Real estate transactions	14,767	13,844	14,472	15,349	16,704	17,867
Professional, scientific and technical activities	59,181	56,886	57,812	60,324	63,350	66,739
Administrative and support service activities	19,480	19,406	19,965	20,802	22,285	22,848
Education	2681	3772	4252	4942	5811	6393
Health and social work	8677	10,093	10,959	13,188	15,251	17,114
Entertainment, cultural and recreational activities	4990	5758	6778	7740	9003	9945
Other service activities	11,772	11,150	11,790	12,786	14,317	15,494

Table A.2 The situation of employees by fields of activity in the period 2015–2020

Economic activities	Year					
	2015	2016	2017	2018	2019	2020
	Number	Number	Number	Number	Number	Number
Total	5,046,317	4,611,395	4,759,419	4,945,868	5,068,063	5,164,471
Agriculture, forestry and fishing	104,819	112,699	117,046	121,720	123,821	126,554
Extractive industry	1,605,611	1,334,943	1,352,862	1,400,975	1,409,137	1,398,710
Manufacturing industry	83,549	55,445	54,234	52,600	52,672	52,824
Production and supply of electricity and heat, gas, hot water and air conditioning	99,453	99,314	99,605	101,810	102,311	104,594
Water distribution; sanitation, waste management, decontamination activities	457,895	354,706	365,298	370,415	374,966	395,669
Construction	848,646	767,525	799,735	833,932	861,875	887,387
Wholesale and retail trade; repair of motor vehicles and motorcycles	284,664	256,480	264,682	276,909	283,853	290,019
Transport and storage	118,306	133,848	151,230	169,837	180,218	187,057
Hotels and restaurants	121,421	143,274	154,520	170,274	182,282	193,955
Information and communications	107,717	88,421	90,160	88,591	90,519	88,945
Financial intermediation and insurance	31,323	24,357	27,448	27,747	29,102	29,870
Real estate transactions	133,022	143,863	151,242	158,459	171,756	179,806
Professional, scientific and technical activities	195,196	259,683	278,540	290,120	300,465	304,768
Administrative and support service activities	213,199	194,087	195,967	200,038	203,050	206,236
Education	393,724	357,125	352,742	352,153	350,812	349,664

(continued)

Table A.2 (continued)

Economic activities	Year					
	2015	2016	2017	2018	2019	2020
	Number	Number	Number	Number	Number	Number
Health and social work	350,278	331,792	345,501	367,231	381,152	395,462
Entertainment, cultural and recreational activities	40,032	62,692	64,450	68,041	72,347	74,732
Other service activities	40,464	45,900	47,996	49,426	52,708	55,637

References

Abd Rashid AF, Yusoff S (2015) A review of life cycle assessment method for building industry. Renew Sustain Energy Rev. https://doi.org/10.1016/j.rser.2015.01.043

Adler M, Dumas B (1984) Exposure to currency risk: definition and measurement. Financ Manage. https://doi.org/10.2307/3665446

Aloini D, Dulmin R, Mininno V (2012) Risk assessment in ERP projects. Inf Syst. https://doi.org/10.1016/j.is.2011.10.001

Arena M, Arnaboldi M, Azzone G (2010) The organizational dynamics of Enterprise Risk Management. Acc Organ Soc. https://doi.org/10.1016/j.aos.2010.07.003

Aven T (2016) Risk assessment and risk management: Review of recent advances on their foundation. Eur J Oper Res. https://doi.org/10.1016/j.ejor.2015.12.023

Василев Д. Ж. Д. (2014) Risk assessment in human health through interactive online tools OIRA. Int Sci Conf UNITECH

Chander DS, Cavatorta MP (2017) An observational method for Postural Ergonomic Risk Assessment (PERA). Int J Ind Ergon. https://doi.org/10.1016/j.ergon.2016.11.007

Dahl MS (2011) Organizational change and employee stress. Manage Sci. https://doi.org/10.1287/mnsc.1100.1273

Dionne G (2013) Risk management: History, definition, and critique. Risk Manag Insur Rev. https://doi.org/10.1111/rmir.12016

Dudley SE, Hays SL, OMB OIRA (2007) Updated principles for risk analysis: memorandum for the heads of executive departments and agencies, Office of Mgmt. and Budget (OMB)

Eaton C (2015) Enterprise risk management. In: Canadian Nuclear Society—35th Annual Conference of the Canadian Nuclear Society and 38th CNS/CNA Student Conference 2015. https://doi.org/10.4018/ijrcm.2014040102

Fabbri J (2020) Risk scoring methodology—RiskWatch. Risk Watch

Fan Y, Stevenson M (2018) A review of supply chain risk management: definition, theory, and research agenda. Int J Phys Distrib Logist Manag. https://doi.org/10.1108/IJPDLM-01-2017-0043

Fischhoff B, Watson SR, Hope C (1984) Defining risk. Policy Sci. https://doi.org/10.1007/BF00146924

Gaureanu A et al (2016) Knowledge management and ergonomics integration for occupational risks management. In: Proceedings of the 28th International business information management association conference—Vision 2020: innovation management, development sustainability, and competitive economic growth

Heckmann I, Comes T, Nickel S (2015) A critical review on supply chain risk—Definition, measure and modeling. Omega (united Kingdom). https://doi.org/10.1016/j.omega.2014.10.004
Institutul Național de Statistică (2019) *INSSE - Baze de date statistice - TEMPO-Online serii de timp, website*
ISO—The International Organization for Standardization (2009) ISO 31000:2009—Risk management—Principles and guidelines. ISO 31000:2009
(IRM), I. of R. M. (2010) A structured approach to Enterprise Risk Management (ERM) and the requirements of ISO 31000 Contents. Risk Manage. https://doi.org/10.1016/j.solmat.2010.12.013
Ivascu VL, Cirjaliu B, Draghici A (2015) Operationalization of the professional risks assessment activity. ACTA Universitatis Cibiniensis. https://doi.org/10.1515/aucts-2015-0030
Izvercian M, Ivascu L (2014) Semantic assessment of the risk: a multidimensional data model for sustainable enterprise. Procedia Technol. https://doi.org/10.1016/j.protcy.2014.10.008
Lalonde C, Boiral O (2012) Managing risks through ISO 31000: a critical analysis. Risk Manage. https://doi.org/10.1057/rm.2012.9
Latham SF, Braun M (2009) Managerial risk, innovation, and organizational decline. J Manag. https://doi.org/10.1177/0149206308321549
Lefèvre S, Vasquez D, Laugier C (2014) A survey on motion prediction and risk assessment for intelligent vehicles. ROBOMECH J https://doi.org/10.1186/s40648-014-0001-z
Luko SN (2013) Risk management principles and guidelines. Qual Eng. https://doi.org/10.1080/08982112.2013.814508
Management R, Guides T (2014) Hazard & Operability Analysis (HAZOP) 1 Overview. Risk Manage Training Guides
Mayer RC, Davis JH, Schoorman FD (1995) An integrative model of organizational trust. Acad Manag Rev https://doi.org/10.5465/amr.1995.9508080335
McCracken J (2003) Oxford dictionary of English. https://doi.org/10.3115/1067737.1067764
Oxford English Dictionary (2017) Oxford English Dictionary Online, Oxford English Dictionary
Rass S et al (2020) Risk management. In: Advanced sciences and technologies for security applications. https://doi.org/10.1007/978-3-030-46908-5_6
Rausand M (2013) Risk assessment: theory, methods, and applications. https://doi.org/10.1002/9781118281116
Rivero LC et al (2015) Fuzzy logic and RULA method for assessing the risk of working. Procedia Manufact. https://doi.org/10.1016/j.promfg.2015.07.591
SheikAllavudeen S, Sankar SP (2015) Hazard identification, risk assessment and risk control. SSRG Int J Ind Eng (SSRG - IJ IE)
Słowikowski M et al (2017) Selected aspects of implementation of "EU occupational safety and health (OSH) strategic framework 2014–2020" connected with automation & robotics. In: Advances in Intelligent Systems and Computing. https://doi.org/10.1007/978-3-319-48923-0_33
Soin K, Collier P (2013) Risk and risk management in management accounting and control. Manag Account Res. https://doi.org/10.1016/j.mar.2013.04.003
Šotić A, Rajić R (2015) The review of the definition of risk. Online J Appl Knowl Manage
Tang CS (2006) Perspectives in supply chain risk management. Int J Prod Econ. https://doi.org/10.1016/j.ijpe.2005.12.006
Teece D, Peteraf M, Leih S (2016) Dynamic capabilities and organizational agility: risk, uncertainty, and strategy in the innovation economy. Calif Manage Rev. https://doi.org/10.1525/cmr.2016.58.4.13
The Institute of Risk Management (2002) A risk management standard. Alarm, Irm
Theoharidou M, Mylonas A, Gritzalis D (2012) A risk assessment method for smartphones. In: IFIP Advances in Information and Communication Technology. https://doi.org/10.1007/978-3-642-30436-1_36
Vassilev D, Dimitrova G (2015) APPLICATION OF INTERACTIVE ONLINE TOOLS OiRA FOR RISK ASSESSMENT OF WASTEWATER TREATMENT. Int J Mach Technol Mater
Van Der Walt J (2003) Risk. In: Hatch and Sumner's Textbook of Paediatric Anaesthesia Third Edition. https://doi.org/10.5949/upo9781846317729.004.
Wohl DL, Harvey L (005) Risk management. In: Complications in Pediatric Otolaryngology. https://doi.org/10.5005/jp/books/11259_10.

Chapter 2
An Integrated Bayesian BWM and Classifiable TOPSIS Model for Risk Assessment

Huai-Wei Lo and James J. H. Liou

Abstract Due to the constant occurrence of natural disasters and human error, risk assessment has become one of the indispensable tasks for governments, organizations, enterprises, etc. In recent years, the risk assessment model based on multiple criteria decision-making (MCDM) is quite popular. Methodologies of this type must rely on experts to assist decision-making in order to make risk analysis results more reliable. However, best-worst method (BWM) is based on pairwise comparison to determine the weight method, which overcomes many shortcomings of analytic hierarchy process (AHP). Currently, BWM has been widely used in various risk management and decision-making issues. In this study, we propose an integrated Bayesian BWM and classifiable technique for order preference by similarity ideal solution (classifiable TOPSIS) model to rank critical failure modes. First, Bayesian BWM is used to generate the group weights of risk factors. Bayesian BWM optimizes original BWM, which effectively integrates the judgments of multiple experts. Then, classifiable TOPSIS is used to rank and classify failure modes. The feasibility of the proposed model was demonstrated by conducting a case study involving a computer numerical control (CNC) rotary table. The analysis results showed that the model can effectively help risk analysts in assessing the risk level of failure modes.

Keywords Risk assessment · MCDM · BWM · TOPSIS

2.1 Introduction

Risk assessment has been regarded as one of the scientific fields for about 40 years. Initially, how to conceptualize risk assessment was a question discussed by researchers at that time. After that, many concepts, principles, theories, methods,

---

H.-W. Lo
Department of Business Administration, Chaoyang University of Technology, Taichung, Taiwan

J. J. H. Liou (✉)
Department of Industrial Engineering and Management, National Taipei University of Technology, Taipei, Taiwan

J. Ren (ed.), *Multi-Criteria Decision Analysis for Risk Assessment and Management*, Industrial Ecology and Environmental Management 1,
https://doi.org/10.1007/978-3-030-78152-1_2

and frameworks for assessing risk have been gradually developed. These studies still deeply influence the field of risk assessment (Aven 2016). Many novel risk assessment methods have brought many contributions and new trends in this field. Newer, more effective, more comprehensive, and more systematic assessment models are the common goal of researchers. A risk decision model is proposed by Hansson and Aven (2014), the model divides risk assessment into five stages, including evidence, knowledge base, broad risk evaluation, decision-makers review, and decision. The first three stages are fact-based, providing evidence through testing and collecting data or information about risk events. Related expert groups are based on these data or information to further research and analyze risks. In addition, the last three stages are value-based. Because risk events are complex and difficult to fully illustrate and explain through scientific tools, after extensive risk assessment, it is necessary to review and judge by decision-makers before final decisions can be made. The purpose of risk analysis is to eliminate and control potential risk events or hazardous factors, so as to reduce the occurrence of accidents and diminish the severity of accidents (Lo et al. 2019).

Research on risk analysis can be divided into three types, including qualitative analysis, semi-quantitative analysis, and quantitative analysis (Marhavilas et al. 2011; Mutlu and Altuntas 2019). Qualitative analysis is to explore whether the research subjects have special attributes or characteristics, and whether they are related or not, through observation and analysis experiments. The data recorded through expert interviews is a way of qualitative analysis. Common qualitative analysis techniques are checklist, what if analysis, safety audits, task analysis, sequentially timed events plotting (STEP), human factors analysis and classification system (HFACS), and hazard and operability study (HAZOP). Quantitative analysis is the quantitative relationship among the components contained in a research object, or the quantitative relationship among the characteristics that it possesses, and it can also analyze and compare the special relationships, characteristics, and properties of several objects quantitatively simultaneously. Therefore, the analysis results are mostly described and solved in terms of quantity. Quantitative analysis techniques include clinical risk and error analysis (CREA), proportional risk-assessment (PRAT), decision matrix risk-assessment (DMRA), societal risk, predictive epistemic approach (PEA), and quantitative risk-assessment (QRA). The semi-quantitative analysis is somewhere in between. In the risk assessment, the collected observation data and survey information use semi-quantitative analysis to perform risk assessment. Semi-qualitative quantitative analysis techniques include event tree analysis (ETA), failure mode and effects analysis (FMEA), fault tree analysis (FTA), risk-based maintenance (RBM), and human error analysis techniques (HEAT/HFEA). The above-mentioned risk assessment and analysis tools are popular methods used by academic researchers and risk analysts today (Marhavilas et al. 2011; Lo et al. 2019; Mutlu and Altuntas 2019; Chang et al. 2019; Gul et al. 2020; Yucesan and Gul 2020; Lo et al. 2020; Liou et al. 2020).

In recent years, many researchers have pointed out that FMEA is one of the most powerful risk assessment tools (Chang et al. 2019; Gul et al. 2020; Yucesan and Gul 2020; Lo et al. 2020; Liou et al. 2020). FMEA is a forward-looking and systematic risk assessment theory used to identify and detect potential risks of incidents. It was first proposed by the US military in 1940 and was used to assess the reliability of the weapons and military systems of US military. Finally, in 1949, the military system standard MIL-STD-1629 was officially published. However, this standard did not fully meet the strict requirements of the US military and the expectations of management. Therefore, a revised version was released in 1980, namely, MIL-STD-1629A (US Department of Defense Washington 1980). In addition, the national aeronautics and space administration (NASA) applied FMEA in the Apollo space program during 1960. FMEA effectively assisted in analyzing the key factors of potential failure of space missions. In 1985, the international electrotechnical commission (IEC) proposed the FMEA specification, called IEC 60812. Since then, FMEA has been widely used in the risk assessment of products or systems in various industries (International Electrotechnical Commission 1985). In 1993, the automotive industry used FMEA to assess product risks in the design and manufacturing stages, and summarized failure modes that may affect product quality. Companies such as automotive industry American society for quality (ASQ), action group (AIAG), DaimlerChrysler corporation, General motors corporation, and Ford Motor company have jointly released a set of FMEA operation instruction manuals. The instruction manual is based on the requirements of the quality system QS-9000 (Automotive Industry Action Group 2008). The international organization for standardization (ISO) is a quality assurance organization that sets many quality requirements and specifications for enterprises. It incorporates FMEA as an analysis tool for evaluating production quality standards ISO-9000 series. Today's FMEA has been applied to analyze powerful technologies for the safety and reliability of products and processes in machinery, nuclear energy, automotive, aviation, food, semiconductor, and medical industries. In addition, it is effective to determine the priority of improvement through FMEA evaluation to prevent catastrophic risk accidents. The successful implementation of FMEA can greatly reduce the occurrence of system or product failures and improve the operational robustness of governments, and enterprises (Ghoushchi et al. 2019; Wang et al. 2019; Qin et al. 2020; Lo et al. 2019).

FMEA defines a set of typical risk factors as evaluation criteria, which are severity (S), occurrence (O), and detectability (D). Taking manufacturing systems/products/equipment as an example, severity is to evaluate the severity of damage to the entire system/product/equipment when a failure mode occurs. A product may have many potential failure modes. In practical application, risk analysts in different industries or different jobs will not have the same considerations, and the methods of evaluation levels used will also be different. Generally, the higher the level, the more serious the damage caused by the failure mode (Lo and Liou 2018). Occurrence is a consideration of the frequency of occurrence of failure modes. This value can be obtained by checking the related failures of similar systems/products/equipment that have been recorded. The higher the rating, the higher the probability of the failure mode occurring (Tooranloo and Ayatollah 2017). Detectability refers to the possibility that failure modes can be identified or discovered. Before the failure mode occurs, it is effective and early to predict when the failure mode may occur. If the detectability is high, it means that the risk of failure can be easily detected early; if the detectability is low, it means that the detectability of the failure is low, and the resulting risk is also high (Chang et al. 2019). Conventional FMEA uses a scale of 1–10 to rate failure modes. Since there are three risk factors, each failure mode will receive three ratings. By calculating the risk priority number (RPN), risk managers can focus on improving the failure mode with the highest RPN instead of focusing on all failure modes. RPN is to multiply the scores of failure modes under the three risk factors of S, O, and D. However, the minimum RPN is 1, and the maximum can reach 1000. Many current FMEA studies believe that there are many shortcomings in judging the degree of risk by RPN, which are described as follows (Chang 2016; Ahn et al. 2017; Ghoushchi et al. 2019; Hu et al. 2019; Wang et al. 2019; Qin et al. 2020; Gul et al. 2020; Lo et al. 2020).

(i) The weights of the risk factors are not considered;
(ii) Different combinations of three risk factors may form the same RPN, but it does not mean that the implication of risk is the same;
(iii) The calculation of RPN using the multiplication of S, O, and D is too simple, and the assessment model is not robust;
(iv) The RPN calculation equation is very sensitive to changes in risk factors;
(v) There are other factors that have not been considered;
(vi) The method of integrating the opinions of multiple risk analysts is too simple, which may easily cause information loss;
(vii) The interdependence of risk factors and failure modes has not been considered.

Many studies have combined FMEA with the multiple criteria decision-making (MCDM) model to enhance the applicability of the risk assessment model and overcome these defects. MCDM converts the information from expert interviews into calculable quantitative data, and then simplifies, integrates, and analyzes to obtain easier-to-interpret information to assist decision-makers in formulating the most appropriate strategy (Lo et al. 2020). MCDM is a decision-making tool based on soft computing, which has excellent analysis capability in a complex assessment environment. When faced with things that have not happened yet, it is difficult to use accurate quantitative data from the past to construct decision-making models. It is feasible to construct a systematic assessment framework through the senior experience of practitioners/risk analysts (Chang et al. 2019; Liou et al. 2020).

This paper proposes an integrated Bayesian best-worst method (Bayesian BWM) (Mohammadi and Rezaei 2020) and a classifiable technique for order preference by similarity ideal solution (TOPSIS) (Liaw et al. 2020) to identify and classify failure modes, which provides an effective and reliable risk assessment model for risk managers and decision-makers, and further supports the development of improvement and preventive measures. This study can be divided into three stages. First, based on the FMEA, several risk factors were determined, namely, S, O, D, and E ("expected cost" added in this study). Bayesian BWM is applied to obtain the importance weights of risk factors. The original BWM overcomes the shortcomings of analytic hierarchy process (AHP). It uses fewer pairwise comparisons to obtain more reliable results. The BWM that joins Bayesian statistical theory can effectively integrate the opinions of multiple risk analysts to obtain a set of group weights. Then, all the failure modes are determined by risk analysts in the industry, and then further screening is made to obtain the final potential failure modes. Risk analysts are invited to assess these failure modes, that is, their relative degrees of risk under each risk factor. In this paper, Bayesian BWM was also used to perform pairwise comparisons to obtain failure mode scores. This approach can prevent risk analysts from getting into a difficult-to-evaluate dilemma when assessing the risk of individual failure modes without a benchmark for comparison. Finally, the classifiable TOPSIS was used to integrate the final risk scores of the failure modes and prioritize their improvement. The classifiable TOPSIS improves the shortcomings of RPN calculation and improves the robustness of analysis results. The calculation process of the proposed model is rigorous and reasonable. We take products from the machine tool manufacturing industry to illustrate the usefulness and effectiveness of the proposed model. This research has made some contributions and improvements in the field of risk assessment, including the following:

(i) The proposed evaluation model takes the cost factor into consideration, and it adds expected cost (E) as a risk factor.
(ii) Bayesian BWM is used to determine the weights of risk factors. In addition, it is also used to measure the relative degrees of risks of failure modes.
(iii) This study applies classifiable TOPSIS, which divides the score of failure modes into four risk levels.
(iv) The introduction of BWM into the field of risk assessment makes it easier for researchers and risk managers to understand the application of this method.
(v) A novel risk assessment model has been introduced in the machine tool industry to assist the industry in formulating decision-making guidelines for risk management.

2.2 Literature Review

2.2.1 FMEA and MCDM

The FMEA program commences with a review of design details, illustrations of equipment block diagrams, and recognition of all possible failures, consecutively. Following recognition, all potential causes and effects should be classified to the related failure modes. After this, failure modes are prioritized based on their destructive effects and ranked by a risk rating (Lo and Liou 2018). More methods are being combined with FMEA to optimize the risk assessment model, with MCDM being the most outstanding (Huang et al. 2020). The relevant MCDM–based FMEA model in the past five years is presented in Table 2.1.

The above-mentioned literatures have made great contributions to the field of risk assessment, making semi-quantitative analysis more effective. After extensive FMEA literature review, this study found some research gaps. As mentioned earlier, FMEA

Table 2.1 Literature review of FMEA model

Authors (year)	MCDM method	Application
Chang (2016)	AHP and 2–tuple representation method	TFT–LCD product troubleshooting
Chai et al. (2016)	Interval type–2 fuzzy sets (IT2FSs)	Analysis of edible bird nest farming
Safari et al. (2016)	Fuzzy VlseKriterijumska Optimizcija I Kaompromisno Resenje in Serbian (VIKOR)	Identifying and evaluating enterprise architecture risks
Wang et al. (2016)	Interval–valued intuitionistic fuzzy sets, complex proportional assessment (COPRAS) and analytic network process (ANP)	Analysis of the hospital service setting
Tooranloo and Ayatollah (2017)	Intuitionistic fuzzy logic	Evaluation the failure modes for the quality of internet banking services
Chen (2017)	Decision-making trial and evaluation laboratory (DEMATEL) and ANP	Prioritization of corrective actions from utility viewpoint
Ahn et al. (2017)	Fuzzy theory	Risk assessment in a hybrid molten carbonate fuel cell and gas turbine system
Mohsen and Fereshteh (2017)	VIKOR and entropy	Risk assessment of the geothermal power plant (GPP)
Nazeri and Naderikia (2017)	Fuzzy ANP and DEMATEL	Identification the risk of failure for railway tamping equipment in Iran
Zhao et al. (2017)	Multi-objective optimization on the basis of ratio analysis (MULTIMOORA), entropy and interval–valued intuitionistic fuzzy set	Risk assessment and system safety analysis
Lo and Liou (2018)	BWM and probability–based grey relational analysis (GRA)	Smartphone failure analysis
Hu et al. (2019)	GRA and TOPSIS	A healthcare risk analysis about suctioning by endotracheal tube (ETT)
Chang et al. (2019)	Rough BWM and rough TOPSIS	Audio equipment failure analysis
Rezaee et al. (2020)	Linguistic FMEA, fuzzy inference system (FIS) and fuzzy data envelopment analysis (DEA)	Producing chemical fertilizers, Sulfuric acid, and other allowable chemicals

(continued)

Table 2.1 (continued)

Authors (year)	MCDM method	Application
Srivastava et al. (2020)	Fuzzy decision support system and fuzzy GRA	Sugar plants milling system
Liou et al. (2020)	Neutrosophic BWM and WASPAS	Switched-Mode Power Supply Risk Analysis

is based on risk analyst judgement to construct a risk assessment matrix. However, in some cases, if risk analysts cannot make appropriate semantic judgments, they can only compare the risk levels from the existing failure modes. In addition, only three risk factors are considered in most RPN calculations, and only a few researches have discussed the management costs of failure modes.

2.2.2 *BWM*

The operation of BWM is easy to understand and to obtain highly consistent results, so it has been widely used in decision-making problems in various industries. BWM plays a very important role in this research. It is not only used to evaluate the weights of risk factors, but also used to evaluate the degrees of risks of failure modes. BWM was proposed by Rezaei (2015), and it is mainly used to overcome the limitations and shortcomings of AHP. Figure 2.1 shows a schematic diagram of the conventional pairwise comparison method. The evaluation system has five evaluated items, and the number of pairwise comparisons is 10 [n $(n-1)/2$ => 5 * (4)/2 = 10]. The pairwise comparison concept proposed by BWM is shown in Fig. 2.2. Experts or decision-makers or risk analysts select the most and least important evaluated items (best and worst evaluated items), and then make pairwise comparisons. Using the same example, BWM only needs 7 pairwise comparisons [$2n-3$ => 2 * (5)–3 = 7].

Tables 2.2, 2.3, and 2.4 illustrate the questionnaire design pattern associated with the AHP and BWM. From the perspective of the experts or decision-makers or risk analysts answering the questionnaires, the BWM questionnaire is more logical and consistent. In Tables 2.2, 2.3, and 2.4, the light gray shaded cells indicate the evaluation information input by the experts. In the AHP example, considering the goal (evaluation issue), a number from 1/9 to 9 is assigned to show the preference of a specific evaluated item over the others (filled in light gray shaded cells). It is worth mentioning that BWM uses a number between 1 and 9 to show the preference of evaluated item i over the evaluated item j.

The steps of BWM in FMEA can be summarized as follows (Rezaei 2015):

Step 1. Construct a set of risk factors.

The experts or decision-makers or risk analysts form an FMEA team, and they formulate evaluation risk factors (C_1, C_2,…, C_j,…, C_n) for decision-making problems.

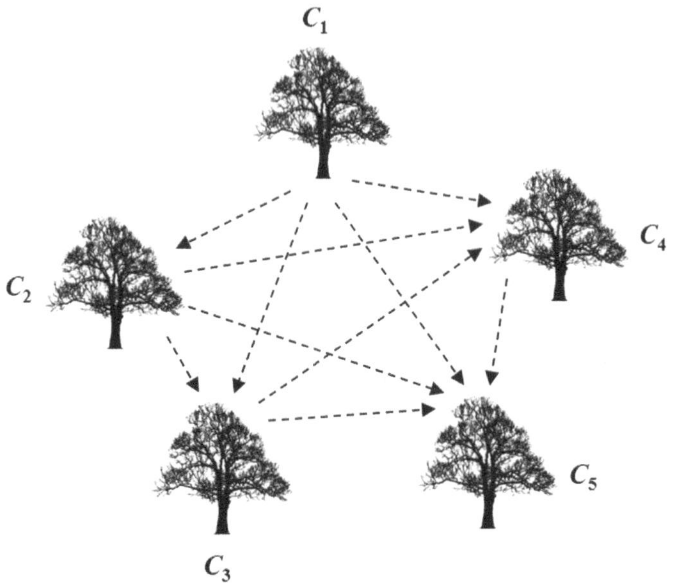

Fig. 2.1 Conventional pairwise comparison method

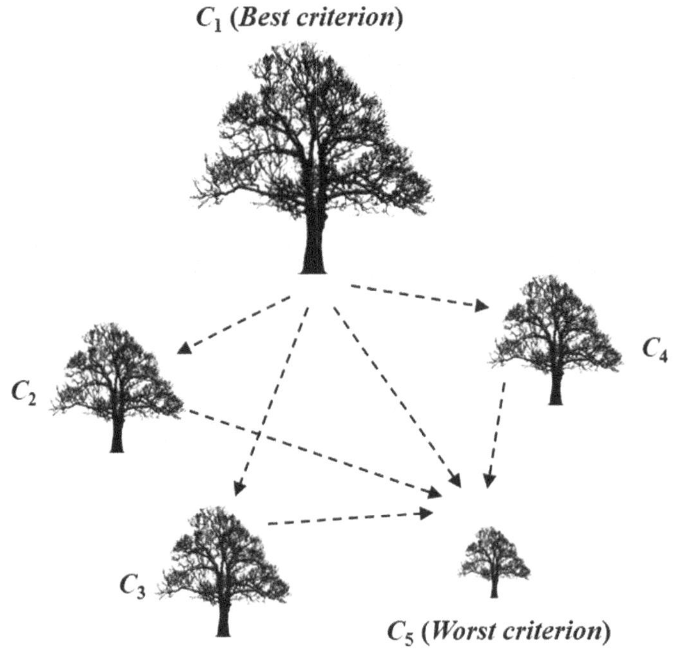

Fig. 2.2 BWM pairwise comparison method

Table 2.2 An example of the AHP questionnaire

	C_1	C_2	C_3	C_4	C_5
C_1					
C_2					
C_3					
C_4					
C_5					

Table 2.3 An example of the BWM questionnaire (best-to-others, BO)

	C_1	C_2	C_3	C_4	C_5
C_1 (Best evaluated item)					

Table 2.4 An example of the BWM questionnaire (others-to-worst, OW)

	C_5 (Worst evaluated item)
C_1	
C_2	
C_3	
C_4	
C_5	

Step 2. Determine the most important (best) and least important (worst) risk factors.

Each expert or decision-maker selects the risk factors from the n risk factors determined in Step 1, that they consider to be the most important (i.e., the best, most superior, or influential) and the least important (i.e., worst, least superior, influential).

Step 3. Obtain the BO vector.

The conventional pairwise comparison way is that all the risk factors need to be compared. Here, the risk analysts only need to compare the best risk factor B with other risk factors j. The scale of evaluation is 1–9, with 1 indicating equal importance, and a larger scale indicating greater importance. The BO vector is as follows:

$$A_B = \left(a_{B1}, a_{B2}, \ldots, a_{Bj}, \ldots, a_{Bn}\right)$$

Step 4. Obtain the OW vector

Similar to step 3, the risk analysts compare the importance of the other risk factors j over the worst risk factor W. The OW vector is as follows:

$$A_W = \left(a_{1W}, a_{2W}, \ldots, a_{jW}, \ldots, a_{nW}\right)^T$$

where a_{BB} and a_{WW} are equal to 1.

Step 5. Construct a linear programming model and obtain the optimal weights.

The optimal risk factors weights w_j are obtained by finding the minimum value in the maximum absolute deviation of $\left|\frac{w_B}{w_j} - a_{Bj}\right|$ and $\left|\frac{w_j}{w_W} - a_{jW}\right|$. Therefore, the min–max model (conceptual model) can be constructed as follows:

$$\min_j \max\left\{\left|\frac{w_B}{w_j} - a_{Bj}\right|, \left|\frac{w_j}{w_W} - a_{jW}\right|\right\}$$

s. t.

$$\sum_j w_j = 1,$$

$$w_j \geq 0, \text{ for all} j. \tag{2.1}$$

Mode (2.1) can be converted into a computational model (2.2) as follows:

min ξ

s. t.

$$\left|\frac{w_B}{w_j} - a_{Bj}\right| \leq \xi, \text{ for all } j,$$

$$\left|\frac{w_j}{w_W} - a_{jW}\right| \leq \xi, \text{ for all } j,$$

$$\sum_j w_j = 1,$$

$$w_j \geq 0, \text{ for all } j. \tag{2.2}$$

Model (2.2) has a total of $4n - 5$ linear constraints, of which there are $2(2n - 3)$ comparison constraints. The ξ value is minimized to converge the solution space to obtain the weights of risk factors $\left(w_1, w_2, \ldots, w_j, \ldots, w_n\right)$. When the solution space of the model (2.2) is larger, the ξ value will be larger, resulting in multiple solutions.

The consistency test of BWM is defined by Rezaei (2015). When a pairwise comparison is fully consistent, it will obey $a_{Bj} \times a_{jW} = a_{BW}$ for all j. For example, a risk analyst evaluates the best risk factor B over the worst risk factor W with a score of 8 (a_{BW}= 8), the best risk factor B over the risk factor C_1 with a score of 4 (a_{B1}= 4), and the risk factor C_1 over the worst risk factor W with a score of 4 (a_{1W}= 2). Obviously, this result is fully consistent (as is the concept of AHP), indicating that the ξ value is 0. Table 2.5 lists the maximum values of ξ (consistency index, CI) for different a_{BW}. Considering the consistency index (Table 2.5), the consistency ratio

Table 2.5 Consistency index (Rezaei 2015)

a_{BW}	1	2	3	4	5	6	7	8	9
CI (max ξ)	0.00	0.44	1.00	1.63	2.30	3.00	3.73	4.47	5.23

(CR) can be calculated as follows:

$$CR = \frac{\xi}{CI} \tag{2.3}$$

The value range of CR is between 0 and 1. The smaller the value, the more consistent it is.

In order to overcome the problem that the conventional BWM has multiple solutions, Rezaei (2016) proposes a linear model of BWM to generate a set of single optimal evaluated items weights (for details, please refer to the Rezaei 2016). The linear programming model (2.4) is as follows:

$\min \xi^L$

$s.\ t.$

$$\left|w_B - a_{Bj}w_j\right| \leq \xi^L, \text{ for all } j,$$

$$\left|w_j - a_{jW}w_W\right| \leq \xi^L, \text{ for all } j,$$

$$\sum_j w_j = 1,$$

$$w_j \geq 0, \text{ for all } j. \tag{2.4}$$

Research on risk assessment through BWM has become increasingly common. For example, Wang and Jin (2019) developed an evaluation model to assess the risks of diversified project financing of city investment companies in China. The structure of projects, structure of financing, asset-liability ratio, earnings before interest and tax (EBIT) margin, ratio of long-term debt to short-term debt, and rate of return on capital are evaluation indicators, and their importance is determined through BWM. In order to reduce the risk of supply chain management, Gan et al. (2019) combined fuzzy theory with BWM to select the most suitable supplier to maintain the stability of material supply. Their study shows that considering the uncertainty of the evaluation environment can more widely envelop the risk analysts' opinions. In addition, there are still many studies that use BWM as a tool for pairwise comparisons of risk factors (Lo and Liou 2018; Lo et al. 2019, 2020).

2.3 The Proposed Risk Assessment Model

The execution process of the proposed risk assessment model is shown in Fig. 2.3. In this section, the basic concepts and calculation process of Bayesian BWM are explained. This paper applies Bayesian BWM to generate the weights of risk factors and assess the risk score of failure modes. Next, the implementation steps of the classifiable TOPSIS technique are introduced. This technique can obtain the priority of the failure modes to provide risk manager with a quick understanding of the relative risk level of the failure modes.

2.3.1 *Bayesian BWM*

In order to more objective integrate the opinions of multiple risk analysts, Mohammadi and Rezaei (2020) developed a probabilistic model of BWM, called Bayesian BWM. Because the judgments provided by each decision-maker in BWM are different, the two vectors information would be different (risk analysts may choose

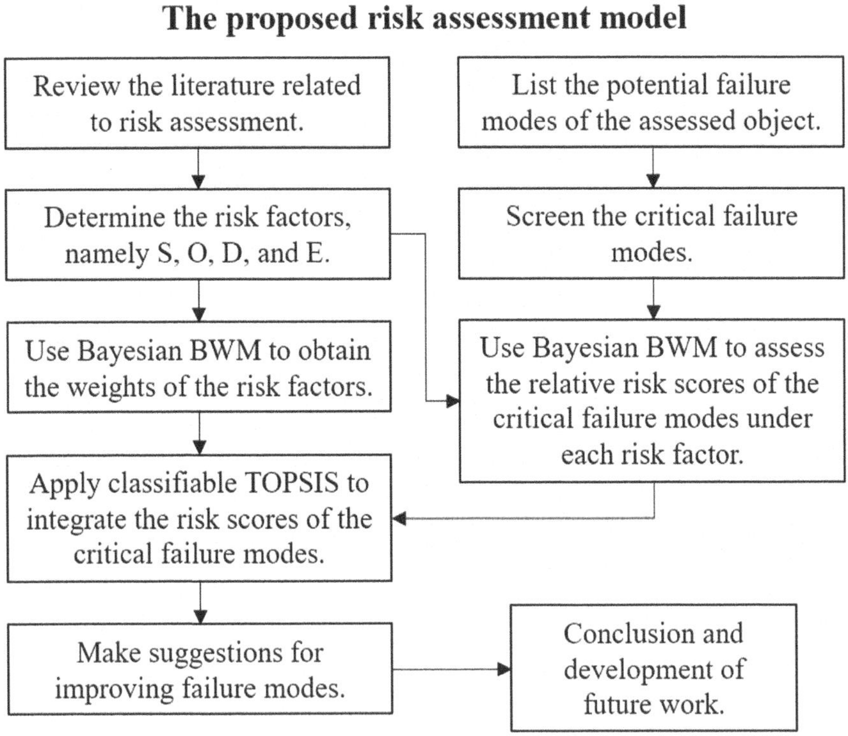

Fig. 2.3 The research flowchart of this study

different best/worst evaluated item). Therefore, it is not possible to aggregate the opinions of multiple risk analysts by using arithmetic averages. The weight vector of the MCDM is $w_j = \left(w_1, w_2, \ldots, w_j, \ldots, w_n\right)$, with $\sum_{j=1}^{n} w_j = 1$ and $w_j \geq 0$ required. Each w_j is represented as the weight of the corresponding evaluated item C_j. The evaluated item C_j can be regarded as a random event, and the weight w_j is their occurrence probability. According to probability theory, the same is true for $\sum_{j=1}^{n} w_j = 1$ and $w_j \geq 0$. Therefore, it is feasible to construct a probability model from the perspective of decision-making. The execution steps are as follows (Mohammadi and Rezaei 2020):

Steps 1–4. The same as Steps 1–4 of conventional BWM.

Step 5. Calculate the best weight value of the group of evaluated items.

The input data A_B and A_W of the BWM can be builded as a probability model of polynomial distribution. The contents of the two vectors are both positive integers, the probability mass function of the polynomial distribution of a given A_W is

$$P(A_W|w) = \frac{\left(\sum_{j=1}^{n} a_{jW}\right)!}{\prod_{j=1}^{n} a_{jW}!} \prod_{j=1}^{n} w_j^{a_{jW}} \tag{2.5}$$

where w is the polynomial distribution, the probability of event j is proportional to the number of occurrences and the total number of experiments.

$$w_j \propto \frac{a_{jW}}{\sum_{j=1}^{n} a_{jW}}, \forall j = 1, 2, \ldots, n. \tag{2.6}$$

Similarly, the worst evaluated item C_W can be written as follows:

$$w_W \propto \frac{a_{WW}}{\sum_{j=1}^{n} a_{jW}} = \frac{1}{\sum_{j=1}^{n} a_{jW}}. \tag{2.7}$$

Equations (2.6) and (2.7) can be integrated to obtain the following:

$$\frac{w_j}{w_W} \propto a_{jW}, \forall j = 1, 2, \ldots, n \tag{2.8}$$

Besides, A_B is modeled using polynomial distribution. However, the concepts of the generation of A_B and A_W are different. The former is the optimal evaluated item B compared to the other evaluated items j. The larger the evaluation value, the smaller the weight of the compared evaluated item j; the latter refers to other evaluated items. The evaluation item j is compared with the worst evaluated item W. The larger the evaluation value, the greater the weight of the evaluated item j. Therefore, the conversion of the assessment content of A_B into the weight should be reciprocal.

$$A_B \sim multinomial\left(\frac{1}{w}\right) \tag{2.9}$$

Equation (2.9) can be written as follows:

$$\frac{1}{w_j} \propto \frac{a_{Bj}}{\sum_{j=1}^{n} a_{Bj}}, \forall j = 1, 2, \ldots, n. \tag{2.10}$$

Similarly, the best rated project c_B can be written as follows:

$$\frac{1}{w_B} \propto \frac{a_{BB}}{\sum_{j=1}^{n} a_{Bj}} = \frac{1}{\sum_{j=1}^{n} a_{Bj}} \Rightarrow \frac{w_B}{w_j} \propto a_{Bj}, \forall j = 1, 2, \ldots, n. \tag{2.11}$$

It can use statistical inference to obtain the optimal weight w_j. Dirichlet probability distribution is used to construct the model (Forbes et al. 2011), whose function is

$$Dir(w|\alpha) = \frac{1}{B(\alpha)} \prod_{j=1}^{n} w_j^{\alpha_j - 1} \tag{2.12}$$

where α is a vector parameter, and w satisfies the restrictions required by MCDM.

Bayesian BWM uses Bayesian to estimate approximate parameters. First, the Dirichlet probability distribution model is used as the prior probability distribution of the weight vector, where α is set to 1, because this parameter does not affect the prior probability. Then, Bayesian estimation is performed based on the parameter w assigned by Dirichlet, then the posterior probability distribution model is

$$\mu_j = \frac{\alpha_{\text{post},t_j} - 1}{\sum_{j=1}^{n} \alpha_{\text{post},t_j} - n} = \frac{1 + a_{jW} - 1}{\sum_{j=1}^{n} \left(a_{jW} + 1\right) - n} = \frac{a_{jW}}{\sum_{j=1}^{n} a_{jW}} \tag{2.13}$$

where $\alpha_{\text{post}} = \alpha + A_W = 1 + A_W$ and $A_W = \left(a_{jW}\right) = (a_{1W}, a_{2W}, \ldots, a_{nW})$.

The distribution model of posterior probability will provide an accurate maximum likelihood estimator. So far, only A_W is considered to estimate the weight, but BWM must consider both vectors A_B and A_W at the same time and needs to integrate the survey data of multiple risk analysts. The above two problems can be solved by using Bayesian BWM. The steps it performs are as follows:

Step 5.1. Construction of joint probability distribution

Assume there are k decision-makers, $k = 1,2,\ldots, K$; the evaluated items $C_j = C_1, C_2,\ldots, C_n$; and the individual best weight after each decision-maker's evaluation is w^k, then the integrated group weight is w^{agg}. $A_B^{1:K}$ represents the vector of the optimal evaluated item compared to other evaluated items by all risk analysts. Similarly, $A_W^{1:K}$ represents the vector of all risk analysts evaluating the other evaluated item compared to the worst evaluated item. The joint probability distribution of group decision is

$$P\left(w^{agg}, w^{1:K} \middle| A_B^{1:K}, A_W^{1:K}\right). \tag{2.14}$$

Each individual variable can be calculated with the following probability rule (the concept of marginal probability function).

$$P(x) = \sum\nolimits_{y} P(x, y) \tag{2.15}$$

where x and y are any random variables.

Step 5.2. Bayesian hierarchical model development and calculation

The weight w^k of each risk analyst depends on the two sets of vectors A_B and A_W, and the optimal weight w^{agg} of the group depends on the optimal weight w^k of each risk analyst. The calculation logic of the Bayesian hierarchical model is based on an iterative method, which means that the direction values A_B and A_W evaluated by each risk analyst will generate w^k. After new evaluation data are added, the group's best weight w^{agg} will be updated continuously. Based on the above concept, the variables are conditionally independent. The joint probability of the Bayesian model is

$$P\left(w^{agg}, w^{1:K} \middle| A_B^{1:K}, A_W^{1:K}\right) \propto P\left(A_B^{1:K}, A_W^{1:K} \middle| w^{agg}, w^{1:K}\right) P\left(w^{agg}, w^{1:K}\right) \tag{2.16}$$

Equation (2.16) can be further derived as follows:

$$\begin{aligned} &P\left(A_B^{1:K}, A_W^{1:K} \middle| w^{agg}, w^{1:K}\right) P\left(w^{agg}, w^{1:K}\right) \\ &= P\left(w^{agg}\right) \prod_{k=1}^{K} P\left(A_W^k \middle| w^k\right) P\left(A_B^k \middle| w^k\right) P\left(w^k \middle| w^{agg}\right) \end{aligned} \tag{2.17}$$

According to Eq. (2.17), the distribution of $A_B^k \middle| w^k$ and $A_W^k \middle| w^k$ can be defined as

$$\begin{aligned} &A_B^k \middle| w^k \sim multinomial\left(\frac{1}{w^k}\right), \forall_k = 1, 2, \ldots, K; \\ &A_W^k \middle| w^k \sim multinomial\left(w^k\right), \forall_k = 1, 2, \ldots, K \end{aligned} \tag{2.18}$$

And w^k can be constructed as Dirichlet distribution under the condition of w^{agg}

$$w^k \middle| w^{agg} \sim Dir\left(\gamma \times w^{agg}\right), \forall_k = 1, 2, \ldots, K \tag{2.19}$$

where w^{agg} is the mean of the distribution, and it is a non-negative parameter.

According to Eq. (2.19), each risk analyst weight w^k will approximate the w^{agg} to the mean of the probability distribution, and the degree of approximation is determined by the parameter. It is reasonable that the parameter distribution obeys the gamma distribution because it has a non-negative limit.

$$\gamma \sim gamma(a, b) \tag{2.20}$$

where a and b are the shape and scale parameters of the gamma distribution.

Finally, the group's optimal weight w^{agg} obeys Dirichlet distribution, and the parameter α is set to 1.

$$w^{agg} \sim Dir(\alpha) \tag{2.21}$$

The calculation the posterior distribution is through Markov-chain Monte Carlo (MCMC) technique. Therefore, the w^{agg} can be obtained according to the above calculation process, which only invites each risk analyst to provide BO and OW vectors.

Step 5.3. Ranking confidence test

Suppose that a group of evaluated items $C_j = (C_1, C_2, \ldots, C_n)$ are evaluated, and two evaluated items are C_i and C_j. It is necessary to confirm whether the ranking results of group weights are consistent with all decision-makers' assessments. Therefore, the concept of Credal Ranking is used to examine the confidence. Then the probability that C_i is better than C_j is

$$P(C_i > C_j) = \int I\left(w_i^{agg} > w_j^{agg}\right)P\left(w^{agg}\right) \tag{2.22}$$

where $P(w^{agg})$ is the posterior probability of w^{agg}, and I is the conditional parameter, which can be calculated only when $\left(w_i^{agg} > w_j^{agg}\right)$ is true, otherwise it is 0.

The confidence level is calculated with the sample size Q obtained by MCMC.

$$P(C_i > C_j) = \frac{1}{Q}\sum_{q=1}^{Q} I\left(w_i^{agg_q} > w_j^{agg_q}\right);$$

$$P(C_j > C_i) = \frac{1}{Q}\sum_{q=1}^{Q} I\left(w_j^{agg_q} > w_i^{agg_q}\right) \tag{2.23}$$

where w^{agg_q} represents w^{agg} of q from the MCMC sample. When $P(C_i > C_j) > 0.5$, it means that the evaluated item i is more important than the evaluated item j, and the probability presented is the confidence level. In addition, the sum of the probabilities is 1, $P(C_i > C_j) + P(C_j > C_i) = 1$.

2.3.2 Classifiable TOPSIS Technique

TOPSIS technique is one of the most popular sorting methods for ranking the evaluated items. This method is to determine the relative position of each evaluated item by determining the degree of separation between each evaluated item and the positive and negative ideal solutions (PIS and NIS). The optimal evaluated item is the one closest to the PIS and the farthest away from the NIS. In risk management, the closer to the positive ideal solution, the greater the degree of risk. TOPSIS will not affect the time and quality of the solution due to the number of evaluated items. In addition, this paper applies classifiable TOPSIS technique (Liaw et al. 2020), which can not only obtain a more reliable ranking, but also divide all the evaluated items into four risk levels. When a new evaluated item is added, the method can be used to immediately assign a level to it. The detailed classifiable TOPSIS technique steps are described as follows (Liaw et al. 2020):

Step 1. Build the initial evaluation matrix ***X***

Assume that there are i evaluated items in the risk assessment framework, $i = 1, 2, \ldots, m$; j represents 4 risk factors, S, O, D, and E. Under each risk factor, the risk values of the evaluated items are evaluated to obtain the initial evaluation matrix. In the paper, Bayesian BWM is used to obtain the content of the matrix.

$$\mathcal{X} = \begin{bmatrix} d_{1S} & d_{1O} & d_{1D} & d_{1E} \\ d_{2S} & d_{2O} & d_{2D} & d_{2E} \\ \vdots & \vdots & \vdots & \vdots \\ d_{iS} & d_{iO} & d_{iD} & d_{iE} \\ \vdots & \vdots & \vdots & \vdots \\ d_{mS} & d_{mO} & d_{mD} & d_{mE} \end{bmatrix}, i = 1, 2, \ldots, m. \tag{2.24}$$

Step 2. Calculate the normalized evaluation matrix $\boldsymbol{X}^*$

Because the data range obtained through Bayesian BWM is already between 0 and 1. Therefore, this step does not need to be executed.

Step 3. Obtain the weighted normalized evaluation matrix $\boldsymbol{X}^{**}$

Taking into account the difference among the importance of the risk factors, the weights of S, O, D, and E are multiplied by the matrix $\boldsymbol{X}^*$ to obtain the weighted normalized evaluation matrix.

$$\mathcal{X}^{**} = \begin{bmatrix} d_{1S} \cdot w_S & d_{1O} \cdot w_O & d_{1D} \cdot w_D & d_{1E} \cdot w_E \\ d_{2S} \cdot w_S & d_{2O} \cdot w_O & d_{2D} \cdot w_D & d_{2E} \cdot w_E \\ \vdots & \vdots & \vdots & \vdots \\ d_{iS} \cdot w_S & d_{iO} \cdot w_O & d_{iD} \cdot w_D & d_{iE} \cdot w_E \\ \vdots & \vdots & \vdots & \vdots \\ d_{mS} \cdot w_S & d_{mO} \cdot w_O & d_{mD} \cdot w_D & d_{mE} \cdot w_E \end{bmatrix}. \tag{2.25}$$

Step 4. Define the PIS and NIS of the evaluated items

One of the advantages of TOPSIS technique is that it can put the two parameters into consideration at the same time, namely, PIS and NIS.

$$\begin{aligned} \text{PIS} &= \left(z_S^+, z_O^+, z_D^+, z_E^+\right) \\ &= (\max\{d_{iS} \cdot w_S\}, \max\{d_{iO} \cdot w_O\}, \max\{d_{iD} \cdot w_D\}, \max\{d_{iE} \cdot w_E\}) \end{aligned} \tag{2.26}$$

$$\text{NIS} = \left(z_S^-, z_O^-, z_D^-, z_E^-\right)$$
$$=(\min\{d_{iS}\cdot w_S\}, \min\{d_{iO}\cdot w_O\}, \min\{d_{iD}\cdot w_D\}, \min\{d_{iE}\cdot w_E\}). \tag{2.27}$$

Step 5. Calculate the distances between the evaluated items and PIS and NIS.

The Euclidean distances are used to determine the distances between the evaluated item i and PIS and NIS. Here, PIS and NIS are also regarded as evaluated items, so that the distance between PIS and NIS can be known.

$$S_i^+ = \sqrt{\left(z_S^+ - x_{iS}^{**}\right)^2 + \left(z_O^+ - x_{iO}^{**}\right)^2 + \left(z_D^+ - x_{iD}^{**}\right)^2 + \left(z_E^+ - x_{iE}^{**}\right)^2} \tag{2.28}$$

$$S_i^+ = \sqrt{\left(x_{iS}^{**} - z_S^-\right)^2 + \left(x_{iO}^{**} - z_O^-\right)^2 + \left(x_{iD}^{**} - z_D^-\right)^2 + \left(x_{iE}^{**} - z_E^-\right)^2}. \tag{2.29}$$

Step 6. Calculate the closeness coefficient (CC_i)

A new index CC_i proposed by Kuo (2017) improves many shortcomings of traditional TOPSIS to obtain more reliable results. Besides, it also gives weights to PIS and NIS, w^+ and w^- represent the relative importance of PIS and NIS, respectively, as shown in Eq. (2.30). Since $w^+ + w^- = 1$, therefore, the settings of w^+ and w^- will affect each other. Generally speaking, the decision-makers will set both w^+ and w^- to 0.5.

$$CC_i = \frac{w^+ S_i^-}{\left(\sum_{i=1}^m S_i^-\right) + S_{\text{PIS}}^-} - \frac{w^- S_i^+}{\left(\sum_{i=1}^m S_i^+\right) + S_{\text{NIS}}^+}. \tag{2.30}$$

However, the proposed ranking index has a shortcoming, that is, when the number of evaluated items increases, the CC_i will also decrease, making it difficult for researchers to interpret the value. Therefore, this study further normalizes CC_i to obtain a new ranking index CC_i^*, as shown in Eq. (2.31).

$$CC_i^* = \frac{CC_i - CC_{\text{NIS}}}{CC_{\text{PIS}} - CC_{\text{NIS}}}. \tag{2.31}$$

Step 7. Set the threshold of the classification levels and draw the outsourcer evaluation graph.

When CC_i^* is closer to 1, it means that it is closer to PIS, and it also means that the risk is very high. On the contrary, when CC_i^* is closer to 0, it means closer to NIS, and the risk of these evaluated items is relatively low. Here, according to the nature of CC_i^*, the decision-making team sets the threshold and then classifies the risk into four levels. Table 2.6 shows the risk classification levels.

Table 2.6 Risk classification levels

CC_i	Classification level	Description
$0.9 \leq CC_i^* \leq 1$	A^+	Level A^+ failure modes have risk score close to the highest risk level and are very dangerous failure modes
$0.6 \leq CC_i^* < 0.9$	A	Level A failure modes is relatively dangerous
$0.3 \leq CC_i^* < 0.6$	B	The risk score of Level B failure modes is average, they belong to the slightly dangerous group
$0 \leq CC_i^* < 0.3$	C	The risk score of Level C failure modes is close to the lowest risk level

2.4 A Real-World Numerical Application

2.4.1 Problem Description

The practicality and effectiveness of the developed risk assessment model can be illustrated through a practical case. The reliability and robustness of machine tools are very important to the manufacturing industry, because it is the main production equipment in the manufacturing industry. Quality control engineers or risk analysts must implement risk assessment and improvement plans for new products to reduce the occurrence of product failures. The company in this case is a multinational manufacturer of machine tool parts in Taiwan. The company's machine tool components include computer numerical control (CNC) rotary tables, indexing tables, hydraulic indexing tables, auto-pallet changer with worktable, etc. In the face of a competitive global market, the company must develop products that are more stable, more precise, faster, and more functional. Therefore, the company implements FMEA activities before the launch of various new products.

The FMEA team was composed of senior department heads of the company. There were seven risk analysts from six different departments, including business department, design department, manufacturing department, quality control department, management department, and sales service department. The seven risk analysts had more than 15 years of experience in the machine tool manufacturing industry and have participated in machine tool related international exhibitions for many times. In addition to their professional technical knowledge, they also understood the development trend of machine tools. In the study, the case company used a newly developed computer numerical control (CNC) rotary table as the product of FMEA analysis, which is CNC rotary tilting Table 250 (TRT-250). As an NC controlled 2 axis table, TRT-250 is suitable for larger workloads in 5 axis machining. A one-piece housing structure with a powerful hydraulic clamping system offers a greater clamping torque and high loading capacities. It is also designed for easy installation and alignment. The FMEA team listed all the failure modes and evaluated the key failure modes, as shown in Table 2.7.

It can be seen from Table 2.7 that there are nine critical failure modes. They are the rotating shaft segmentation accuracy exceeding the standard (FM1), the rotating

Table 2.7 TRT-250 failure modes

No.	Category	Failure mode description	Risk analyst							Total
			1	2	3	4	5	6	7	
1	Rotating shaft part	The rotating shaft segmentation accuracy exceeding the standard	*	*		*	*	*	*	6
2		The rotating shaft reproducibility exceeding the standard	*	*	*	*	*			5
3		The positive/negative clearance of the rotating shaft exceeding the standard	*	*	*	*	*		*	6
4		The motor making noise when rotating								0
5		The machine making noise when the rotating shaft rotates			*					1
6		The verticality of the rotating shaft disk surface and the side surface exceeding the tolerance								0
7		The parallelism between the rotating shaft disk surface and the bottom surface exceeding the tolerance					*			1
8		The disk surface fluctuation exceeding the tolerance when the rotating shaft rotates								0
9	Inclined shaft part	The inclined shaft segmentation accuracy exceeding the standard								0
10		The inclined shaft reproducibility exceeding the standard	*	*	*		*	*		5
11		The positive/negative clearance of the inclined shaft exceeding the standard	*	*	*	*	*	*	*	7
12		The motor making noise when rotating								0
13		The machine making noise when the inclined shaft rotates	*		*		*	*	*	5
14		The verticality of the inclined shaft disk surface and the side surface exceeding the tolerance								0
15		The parallelism between the inclined shaft disk surface and the bottom surface exceeding the tolerance					*			1

(continued)

Table 2.7 (continued)

No.	Category	Failure mode description	Risk analyst							Total
			1	2	3	4	5	6	7	
16		The disk surface fluctuation exceeding the tolerance when the inclined shaft rotates								0
17		The height of the inclined shaft center exceeding the tolerance								0
18	Whole	Rusty or flawed disk surface		*						1
19		Paint peeling				*				1
20		Wrong label				*				1
21		Abnormal proximity switch	*	*	*	*	*	*		6
22		Unable to return to the origin								0
23		Abnormal hydraulic locking								0
24		Abnormality during program execution								0
25		Oil leakage from the base			*					1
26		Oil leakage from the disk surface	*		*	*	*	*	*	6
27		Oil leakage from the motor case								0
28		Improper waterproof measures	*	*	*		*	*	*	6

shaft reproducibility exceeding the standard (FM2), the positive/negative clearance of the rotating shaft exceeding the standard (FM3), the inclined shaft reproducibility exceeding the standard (FM4), the positive/negative clearance of the inclined shaft exceeding the standard (FM5), the machine making noise when the inclined shaft rotates (FM6), abnormal proximity switch (FM7), oil leakage from the disk surface (FM8), and improper waterproof measures (FM9). FMEA was performed to further analyze them.

2.4.2 Applies Bayesian BWM to Obtain Risk Factor Weights

As in the steps of Bayesian BWM introduced in Sect. 2.3.1, first, the seven risk analysts were invited to select the most and least important risk factors based on their judgments. By comparing the importance of the most important risk factors with other risk factors by the seven risk analysts, the BO vectors were constructed, as shown in Table 2.8. For example, risk analyst 1 believed that D is the most important factor, and its importance compared to other factors is 2, 4, 1, and 2. Similarly, the OW vectors can be constructed by comparing other factors to the worst factor, as shown in Table 2.9. The Bayesian BWM questionnaires completed by all risk analysts have been checked for consistency to ensure the quality and logic of all questionnaires. Next, the MATLAB software provided by Mohammadi and Rezaei (2020) was used

Table 2.8 Bayesian BWM input data (BO vectors)

	Best	S	O	D	E
Risk analyst 1	*D*	2	4	1	2
Risk analyst 2	*O*	2	1	2	2
Risk analyst 3	*D*	3	4	1	2
Risk analyst 4	*D*	2	3	1	2
Risk analyst 5	*O*	3	1	3	5
Risk analyst 6	*D*	2	2	1	1
Risk analyst 7	*O*	1	1	1	1

Table 2.9 Bayesian BWM input data (OW vectors)

	Risk analyst 1	Risk analyst 2	Risk analyst 3	Risk analyst 4	Risk analyst 5	Risk analyst 6	Risk analyst 7
Worst	*O*	*E*	*O*	*O*	*E*	*O*	*E*
S	2	1	2	2	3	2	1
O	1	2	1	1	5	1	1
D	4	1	4	3	2	2	1
E	2	1	3	2	1	2	1

Table 2.10 The weights of the risk factors

Risk factor	S	O	D	E
Weight	0.2362	0.2142	0.3209	0.2287
Rank	2	4	1	3

to obtain the integrated weights of risk factors, as shown in Table 2.10. The weights of the factors are $w_S = 0.2362$, $w_O = 0.2142$, $w_D = 0.3209$, and $w_E = 0.2287$, and their importance ranking is $D \succ S \succ E \succ O$.

2.4.3 *Using Bayesian BWM to Evaluate the Risk Scores of Failure Modes*

In addition to measuring the importance of risk factors, Bayesian BWM also serves as a risk score assessment tool for failure modes. It uses each risk factor S, O, D, and E as a basis to evaluate the relative importance of failure modes. For example, based on severity (S), FM9 is the most severe, and the severity scores compared to other failure modes are shown in Table 2.11. Next, FM6 is the least serious, and the scores of other failure modes compared to FM6 are shown in Table 2.12. According to this process, the failure modes can be evaluated under the 4 risk factors. The remaining survey data are shown in Tables 2.13, 2.14, 2.15, 2.16, 2.17, and 2.18.

In this step, the seven risk analysts evaluated each failure mode according to different risk factors, and the evaluation method used was based on pairwise comparisons. Bayesian BWM was used to integrate the evaluation data of all risk analysts and generate an initial evaluation matrix, as shown in Table 2.19. In Table 2.19, the sum of each column must be 1, so there is no need for normalization.

Table 2.11 BO vectors evaluated for failure modes based on severity (S)

	Best	FM1	FM2	FM3	FM4	FM5	FM6	FM7	FM8	FM9
Risk analyst 1	*FM9*	2	2	3	3	1	7	1	5	1
Risk analyst 2	*FM8*	4	5	1	5	1	5	2	1	2
Risk analyst 3	*FM4*	1	2	2	1	2	2	5	2	2
Risk analyst 4	*FM4*	1	1	1	1	1	1	3	2	2
Risk analyst 5	*FM4*	1	2	3	1	3	2	3	2	2
Risk analyst 6	*FM9*	2	2	3	3	1	7	1	4	1
Risk analyst 7	*FM8*	4	4	1	5	1	5	2	1	2

Table 2.12 OW vectors evaluated for failure modes based on severity (S)

	Risk analyst 1	Risk analyst 2	Risk analyst 3	Risk analyst 4	Risk analyst 5	Risk analyst 6	Risk analyst 7
Worst	*FM6*	*FM6*	*FM7*	*FM7*	*FM7*	*FM6*	*FM6*
FM1	3	2	5	2	3	3	2
FM2	4	1	4	2	2	4	2
FM3	2	5	3	2	2	2	5
FM4	3	1	5	3	3	3	1
FM5	7	5	3	3	1	7	5
FM6	1	1	3	3	2	1	1
FM7	6	3	1	1	1	6	3
FM8	2	5	2	2	2	2	5
FM9	7	4	3	2	2	7	4

Table 2.13 BO vectors evaluated for failure modes based on occurrence (O)

	Best	FM1	FM2	FM3	FM4	FM5	FM6	FM7	FM8	FM9
Risk analyst 1	*FM5*	7	6	5	7	1	5	2	3	3
Risk analyst 2	*FM5*	5	5	5	5	1	5	7	5	5
Risk analyst 3	*FM5*	4	5	4	4	1	5	5	4	4
Risk analyst 4	*FM5*	5	5	5	5	1	5	5	1	1
Risk analyst 5	*FM9*	5	6	7	7	3	5	7	2	1
Risk analyst 6	*FM9*	4	4	4	4	4	4	5	1	1
Risk analyst 7	*FM8*	5	4	2	2	2	2	2	1	1

Table 2.14 OW vectors evaluated for failure modes based on occurrence (O)

	Risk analyst 1	Risk analyst 2	Risk analyst 3	Risk analyst 4	Risk analyst 5	Risk analyst 6	Risk analyst 7
Worst	*FM1*	*FM7*	*FM7*	*FM7*	*FM7*	*FM7*	*FM1*
FM1	1	2	2	2	3	2	1
FM2	2	2	1	2	2	1	1
FM3	2	2	2	2	2	2	3
FM4	1	2	2	2	2	1	3
FM5	7	7	5	5	3	2	4
FM6	2	2	2	2	2	1	3
FM7	3	1	1	1	1	1	3
FM8	4	2	2	5	5	5	5
FM9	3	2	2	5	7	5	5

Table 2.15 BO vectors evaluated for failure modes based on detectability (D)

	Best	FM1	FM2	FM3	FM4	FM5	FM6	FM7	FM8	FM9
Risk analyst 1	*FM9*	7	8	7	8	7	6	9	2	1
Risk analyst 2	*FM9*	6	6	6	6	6	6	8	3	1
Risk analyst 3	*FM4*	2	3	3	1	2	2	8	2	3
Risk analyst 4	*FM4*	2	3	3	1	2	2	9	3	2
Risk analyst 5	*FM4*	3	2	2	1	2	2	8	1	1
Risk analyst 6	*FM2*	2	1	2	2	2	8	9	2	2
Risk analyst 7	*FM2*	2	1	3	2	1	7	7	2	1

Table 2.16 OW vectors evaluated for failure modes based on detectability (D)

	Risk analyst 1	Risk analyst 2	Risk analyst 3	Risk analyst 4	Risk analyst 5	Risk analyst 6	Risk analyst 7
Worst	*FM7*	*FM7*	*FM7*	*FM7*	*FM7*	*FM7*	*FM7*
FM1	2	2	4	5	3	4	4
FM2	1	2	3	3	4	9	7
FM3	2	2	3	3	4	5	3
FM4	1	2	8	9	8	5	4
FM5	1	2	4	5	4	5	7
FM6	2	2	4	5	4	4	1
FM7	1	1	1	1	1	1	1
FM8	6	3	4	3	8	5	5
FM9	9	8	3	5	8	5	7

Table 2.17 BO vectors evaluated for failure modes based on expected cost (E)

	Best	FM1	FM2	FM3	FM4	FM5	FM6	FM7	FM8	FM9
Risk analyst 1	*FM9*	5	5	5	5	5	2	5	9	1
Risk analyst 2	*FM9*	4	4	4	4	4	3	4	9	1
Risk analyst 3	*FM9*	4	5	4	5	5	2	5	9	1
Risk analyst 4	*FM9*	5	4	5	4	4	3	4	9	1
Risk analyst 5	*FM9*	4	5	5	5	5	1	5	9	1
Risk analyst 6	*FM9*	4	4	5	4	4	1	4	9	1
Risk analyst 7	*FM9*	4	5	4	5	5	2	5	9	1

Table 2.18 OW vectors evaluated for failure modes based on expected cost (E)

	Risk analyst 1	Risk analyst 2	Risk analyst 3	Risk analyst 4	Risk analyst 5	Risk analyst 6	Risk analyst 7
Worst	*FM8*	*FM8*	*FM8*	*FM8*	*FM8*	*FM8*	*FM8*
FM1	2	3	2	3	2	3	3
FM2	3	2	2	3	2	2	2
FM3	2	3	2	2	2	2	2
FM4	3	3	2	2	2	2	2
FM5	2	2	2	2	3	2	3
FM6	5	6	5	5	4	5	6
FM7	2	3	2	3	3	3	2
FM8	1	1	1	1	1	1	1
FM9	9	9	9	9	9	9	9

Table 2.19 Initial evaluation matrix ***X***

	S	O	D	E
FM1	0.111	0.076	0.101	0.085
FM2	0.099	0.074	0.109	0.079
FM3	0.116	0.086	0.093	0.077
FM4	0.097	0.079	0.135	0.079
FM5	0.161	0.194	0.113	0.079
FM6	0.065	0.086	0.079	0.174
FM7	0.102	0.077	0.040	0.082
FM8	0.104	0.160	0.144	0.040
FM9	0.144	0.167	0.186	0.304
Sum	1	1	1	1

2.4.4 Employs Classifiable TOPSIS to Rank Critical Failure Modes

The manufacturing process of machine tools is complicated and it is not easy to evaluate its reliability. It is feasible to diagnose potential critical failure modes of machine tools through FMEA. This study uses the classifiable TOPSIS technique to rank critical failure modes and classify them. For a more detailed introduction and concept of the classifiable TOPSIS technique, the study of Liaw et al. (2020) can be referred to. The weight results of the risk factors from Sect. 2.4.1 can be substituted into the calculation of classifiable TOPSIS, and the weighted normalized matrix can be obtained, as shown in Table 2.20.

In Table 2.20, PIS and NIS (representing maximization and minimization of risks) are:

Table 2.20 Weighted normalized matrix X^{**}

	S	O	D	E
FM1	0.026	0.016	0.032	0.019
FM2	0.023	0.016	0.035	0.018
FM3	0.027	0.018	0.030	0.018
FM4	0.023	0.017	0.043	0.018
FM5	0.038	0.042	0.036	0.018
FM6	0.015	0.018	0.025	0.040
FM7	0.024	0.017	0.013	0.019
FM8	0.025	0.034	0.046	0.009
FM9	0.034	0.036	0.060	0.070
Max	0.038	0.042	0.060	0.070
Min	0.015	0.016	0.013	0.009

PIS = (0.038, 0.042, 0.060, 0.070);
NIS = (0.015, 0.016, 0.013, 0.009).

Next, the distance between the failure mode and PIS (S^{+}) and NIS (S^{-}) can be calculated through Eqs. (2.28) and (2.29). It is certain that the distance between the highest level and PIS must be 0, and similarly, the distance between the worst level and NIS is also 0. The distance between NIS and PIS is 0.084. Table 2.21 shows the analysis results of the classifiable TOPSIS. The top five failure modes in the ranking are improper waterproof measures (FM9), the positive/negative clearance of the inclined shaft exceeding the standard (FM5), oil leakage from the disk surface (FM8), the machine making noise when the inclined shaft rotates (FM6), and the inclined

Table 2.21 Classifiable TOPSIS analysis results

	S^{+}	S^{-}	CC	CC^{*}	Rank	Risk level
FM1	0.064	0.024	−0.022	0.271	6	C
FM2	0.064	0.025	−0.023	0.271	7	C
FM3	0.065	0.022	−0.026	0.250	8	C
FM4	0.061	0.032	−0.010	0.340	5	B
FM5	0.057	0.042	0.006	0.434	2	B
FM6	0.056	0.033	−0.005	0.371	4	B
FM7	0.075	0.013	−0.046	0.135	9	C
FM8	0.064	0.039	−0.004	0.375	3	B
FM9	0.007	0.081	0.096	0.946	1	A^{+}
Max	0.000	0.084	0.106	1		
Min	0.084	0.000	−0.070	0		
Sum	0.596	0.396	0			

shaft reproducibility exceeding the standard (FM4). The closeness coefficient (CC^*) of FM9 is 0.946 as the maximum value, and it is at the highest risk level, indicating that it is a failure mode that needs to be solved and controlled urgently. In general, four failure modes fall into Risk Level B and three into Risk Level C. Therefore, decision-makers should devote all risk management resources to Levels A^+ and B to prevent these failure modes in order to reduce the risk of product failure.

2.5 Discussion and Conclusions

The analysis of the importance of risk factors shows that "detectability" and "severity" are the two most important factors for evaluating the failure modes of machine tools. Generally speaking, when customers use machine tools to produce products, they expect that the possibility of machine tool failure is very small, so they have a stable production process and stable quality. As the development trend of Industry 4.0 has brought considerable impact to the global machine tool industry, machine tools with self-test reliability have become the basic needs of customers. Therefore, many new product tools have the opportunity to install sensors and monitor the stability of the machine through the Internet of Things technology. In addition, from the perspective of business strategy, it is necessary to properly distribute risk management resources, so it is reasonable to include the expected cost as a risk factor. This concept also echoes the research of Chang (2016), Lo and Liou (2018), Lo et al. (2019), Chang et al. (2019), Liou et al. (2020).

There are many risk assessment techniques and tools that can be used to assess the reliability of machine tools, the most famous of which are FTA and reliability engineering. However, these quantitative analyses require actual measurement data as a basis. New products usually show failure after a certain number of years of use. At present, it is difficult to obtain actual failure information. In recent years, the research on failure mode analysis by combining MCDM and FMEA has gradually increased. The research provides a more effective and simpler method, which is a combination of Bayesian BWM and classifiable TOPSIS technique. Bayesian BWM continues to reduce the number of pairwise comparisons and incorporates Bayesian statistical probability model to aggregate the opinions of multiple risk analysts. The classifiable TOPSIS technique is no longer just a tool to rank failure modes, it can also assign risk levels and give suggestions for improvement, as shown in Table 2.21. The model we propose provides a new risk management evaluation framework that overcomes the limitations of the past FMEA methods.

In terms of closeness coefficient (CC^*), improper waterproof measures (FM9) shows the highest value in the failure mode of TRT-250, which is a risk of Level A^+. Because the distance between FM9 and PIS is 0.007, it is very close to PIS. On the other hand, the distance between FM9 and NIS is 0.081, which is far away from NIS. Water resistance has always been a problem for machine tools. Although we now have good waterproof technology, in the long run, the humidity of the environment and the sprinkling of cooling water will cause machine tools to rust, which will affect its lubricity and accuracy. The case company should conduct more preventive designs for FM9 to extend the product life cycle. In addition, FM5, FM8, FM6, and FM4 are B-level risks. They are not as urgent as risk levels A^+ and A. R&D engineers can control these failure modes through changes of parts design and alike. Therefore, the model can not only provide ranking, but also provide strategies for selecting improved failure modes to achieve the required minimum risk level, which is a new contribution. In short, according to the weight of each risk factor, managers can set priorities for designing repair and maintenance plans and allocate appropriate resources to prevent failure modes.

The model provides a framework for obtaining the priority of machine failure modes based on multiple risk analyst judgments. The main advantages of the proposed integrated risk assessment model are as follows. First of all, Bayesian BWM is an effective method to obtain the weights of risk factors. The factor E has been added to reflect actual budget constraints. Secondly, it is feasible to use Bayesian BWM to rank the failure modes, and the relative importance of the failure modes can be identified through pairwise comparisons. Finally, classifiable TOPSIS can be used to rank failure modes. Decision-makers can determine the relative risk levels of all failure modes based on the ranking index. This method provides a new evaluation framework for the risk analyst of machine tool production. The results of this research are very useful for formulating and improving product design plans and failure prevention strategies.

Although this study remedies some shortcomings of the original FMEA method, there are still some limitations that should be addressed. At present, the uncertainty of information and the confidence of risk analyst judgments have not been considered. Future research can combine Z fuzzy logic to reflect simultaneous uncertainty and confidence. Since this evaluation model is of a semi-quantitative type, it is expected that more actual measured data can be used to construct a purely quantitative analysis in the future.

References

Ahn J, Noh Y, Park SH, Choi BI, Chang D (2017) Fuzzy-based failure mode and effect analysis (FMEA) of a hybrid molten carbonate fuel cell (MCFC) and gas turbine system for marine propulsion. J Power Sources 364:226–233

Automotive industry action group (AIAG) (2008) Potential failure mode and effect analysis (FMEA) reference manual. FMEA reference manual 4th edition

Aven T (2016) Risk assessment and risk management: Review of recent advances on their foundation. Eur J Oper Res 253(1):1–13

Chai KC, Jong CH, Tay KM, Lim CP (2016) A perceptual computing-based method to prioritize failure modes in failure mode and effect analysis and its application to edible bird nest farming. Appl Soft Comput 49:734–747

Chang KH (2016) Generalized multi-attribute failure mode analysis. Neurocomputing 175:90–100

Chang TW, Lo HW, Chen KY, Liou JJ (2019) A novel FMEA model based on rough BWM and rough TOPSIS-AL for risk assessment. Mathematics 7(10):874

Chen JK (2017) Prioritization of corrective actions from utility viewpoint in FMEA application. Qual Reliab Eng Int 33(4):883–894
Forbes C, Evans M, Hastings N, Peacock B (2011) Statistical distributions. Wiley
Gan J, Zhong S, Liu S, Yang D (2019) Resilient supplier selection based on fuzzy BWM and GMo-RTOPSIS under supply chain environment. Discret Dyn Nat Soc 2019:2456260
Ghoushchi SJ, Yousefi S, Khazaeili M (2019) An extended FMEA approach based on the Z-MOORA and fuzzy BWM for prioritization of failures. Appl Soft Comput 81:
Gul M, Yucesan M, Celik E (2020) A manufacturing failure mode and effect analysis based on fuzzy and probabilistic risk analysis. Appl Soft Comput 96:
Hansson SO, Aven T (2014) Is risk analysis scientific? Risk Anal 34(7):1173–1183
Hu YP, You XY, Wang L, Liu HC (2019) An integrated approach for failure mode and effect analysis based on uncertain linguistic GRA–TOPSIS method. Soft Comput 23(18):8801–8814
Huang J, You JX, Liu HC, Song MS (2020) Failure mode and effect analysis improvement: a systematic literature review and future research agenda. Reliab Eng Syst Saf 199:
International electrotechnical commission, Geneva (1985) Analysis techniques for system reliability- procedures for failure mode and effect analysis, Geneva. IEC 60812
Kuo T (2017) A modified TOPSIS with a different ranking index. Eur J Oper Res 260(1):152–160
Liaw CF, Hsu WCJ, Lo HW (2020) A hybrid MCDM model to evaluate and classify outsourcing providers in manufacturing. Symmetry 12(12):1962
Liou JJ, Liu PC, Lo HW (2020) A failure mode assessment model based on neutrosophic logic for switched-mode power supply risk analysis. Mathematics 8(12):2145
Lo HW, Liou JJ (2018) A novel multiple-criteria decision-making-based FMEA model for risk assessment. Appl Soft Comput 73:684–696
Lo HW, Liou JJ, Huang CN, Chuang YC (2019) A novel failure mode and effect analysis model for machine tool risk analysis. Reliab Eng Syst Saf 183:173–183
Lo HW, Shiue W, Liou JJ, Tzeng GH (2020) A hybrid MCDM-based FMEA model for identification of critical failure modes in manufacturing. Soft Comput 24:15733–15745
Marhavilas PK, Koulouriotis DE, Mitrakas C (2011) On the development of a new hybrid risk assessment process using occupational accidents' data: application on the Greek Public Electric Power Provider. J Loss Prev Process Ind 24(5):671–687
Mohammadi M, Rezaei J (2020) Bayesian best-worst method: A probabilistic group decision making model. Omega 96:102075
Mohsen O, Fereshteh N (2017) An extended VIKOR method based on entropy measure for the failure modes risk assessment–a case study of the geothermal power plant (GPP). Saf Sci 92:160–172
Mutlu NG, Altuntas S (2019) Risk analysis for occupational safety and health in the textile industry: Integration of FMEA, FTA, and BIFPET methods. Int J Ind Ergon 72:222–240
Nazeri A, Naderikia R (2017) A new fuzzy approach to identify the critical risk factors in maintenance management. Int J Adv Manuf Technol 92(9–12):3749–3783
Qin J, Xi Y, Pedrycz W (2020) Failure mode and effects analysis (FMEA) for risk assessment based on interval type-2 fuzzy evidential reasoning method. Appl Soft Comput 89:
Rezaee MJ, Yousefi S, Eshkevari M, Valipour M, Saberi M (2020) Risk analysis of health, safety and environment in chemical industry integrating linguistic FMEA, fuzzy inference system and fuzzy DEA. Stoch Environ Res Risk Assess 34(1):201–218
Rezaei J (2015) Best-worst multi-criteria decision-making method. Omega 53:49–57
Rezaei J (2016) Best-worst multi-criteria decision-making method: some properties and a linear model. Omega 64:126–130
Safari H, Faraji Z, Majidian S (2016) Identifying and evaluating enterprise architecture risks using FMEA and fuzzy VIKOR. J Intell Manuf 27(2):475–486
Srivastava P, Khanduja D, Ganesan S (2020) Fuzzy methodology application for risk analysis of mechanical system in process industry. Int J Syst Assur Eng Manag 11(2):297–312

Tooranloo HS, Ayatollah AS (2017) Pathology the internet banking service quality using failure mode and effect analysis in interval-valued intuitionistic fuzzy environment. Int J Fuzzy Syst 19(1):109–123

US Department of Defense Washington, DC (1980) Procedures for performing a failure mode effects and criticality analysis. US MIL-STD-1629A

Wang Y, Jin X (2019) Structural risk of diversified project financing of city investment company in China based on the best worst method. Eng Constr Archit, Manag

Wang LE, Liu HC, Quan MY (2016) Evaluating the risk of failure modes with a hybrid MCDM model under interval-valued intuitionistic fuzzy environments. Comput Ind Eng 102:175–185

Wang W, Liu X, Chen X, Qin Y (2019) Risk assessment based on hybrid FMEA framework by considering decision maker's psychological behavior character. Comput Ind Eng 136:516–527

Yucesan M, Gul M (2020) Hospital service quality evaluation: an integrated model based on Pythagorean fuzzy AHP and fuzzy TOPSIS. Soft Comput 24(5):3237–3255

Zhao H, You JX, Liu HC (2017) Failure mode and effect analysis using MULTIMOORA method with continuous weighted entropy under interval-valued intuitionistic fuzzy environment. Soft Comput 21(18):5355–5367

Chapter 3
A Modified Risk Prioritization Approach Using Best–Worst Method

Muhammet Gul, Melih Yucesan, and Erkan Celik

Abstract Manufacturing plants often face many defects during manufacturing processes. It is not possible to eliminate all failures in the process. Therefore, there is a need for a risk assessment methodology, such as Failure Mode and Effect Analysis (FMEA). The risk priority number (RPN) calculation has been criticized in many aspects in the classical FMEA analysis. Some of these criticisms include the inadequacy of the number of risk factors, uncertainties in determining risk factors, and reaching the same RPN number with different risk combinations. To eliminate the disadvantages of the classic RPN calculation of FMEA, a modified approach was proposed in this study. A case study was carried out in a company that makes plastic production by injection molding in Germany. RPN elements of severity (S), occurrence (O), and detection (D) were evaluated by an occupational health and safety expert from the observed company. The weights of these parameters were calculated using Best and Worst method (BWM). Then, six failure modes were evaluated with respect to S, D, and O elements, and their preference values were also calculated via BWM. Modified RPN values of each failure mode were calculated by multiplying the evaluation matrices and weight matrix. Preventive measures were taken in operation for three failure modes with the highest RPN value.

Keywords Risk priority number · FMEA · Best–Worst method · Plastic injection

M. Gul (✉)
Department of Emergency Aid and Disaster Management, Munzur University, 62000 Tunceli, Turkey
e-mail: muhammetgul@munzur.edu.tr

M. Yucesan
Department of Mechanical Engineering, Munzur University, 62000 Tunceli, Turkey

E. Celik
Department of Transportation and Logistics, Istanbul University, Avcılar, 34322 Istanbul, Turkey

J. Ren (ed.), *Multi-Criteria Decision Analysis for Risk Assessment and Management*, Industrial Ecology and Environmental Management 1,
https://doi.org/10.1007/978-3-030-78152-1_3

3.1 Introduction

Production is one of the most critical factors in increasing societies' life quality and ensuring society's continuity. The manufacturing industry takes a large share in the world economy (Cheung et al. 2017). Globalization and the rapid change of business dynamics threaten the sustainability of manufacturers. Manufacturers have to improve their production performance to compete with other companies continually (Kang and Subramaniam 2018; Zhou and ThaiÑ 2016).

Plastics are relatively inexpensive, strong, and highly corrosion-resistant materials with heat and hot insulation properties (Fuentes-Huerta et al. 2018). Plastic injection, one of the most frequently used methods in the production of plastic products, can be defined as the process of injecting plastic heated to a specific temperature into a mold under a certain pressure (Gökler and Boran 2020; Karasu and Salum 2018; Sadeghi 2000). This method is a very popular production method due to its high productivity, low surface roughness, and relatively low cost (Park and Dang 2017).

Production performance is affected by the uncertainty and difficulty of controlling many parameters, such as machine failures and production errors. Machine failures cause production to stop and increase unexpected costs of the business. The production of defective parts causes an increase in direct and indirect costs due to the enterprise's internal or external low quality (Pan et al. 2010). Many methods have been proposed in recent years to reduce uncertainties and analyze failures in enterprises. One of these methods is FMEA (Bhattacharjee et al. 2020). FMEA was first proposed for the aviation industry in the 1960s. FMEA is used extensively to identify, measure, and eliminate possible errors in systems and processes. FMEA is widely used, especially in the automotive, aviation, railway, and nuclear industries, due to its easy use and effective results (Li et al. 2020; Wang et al. 2018). In the FMEA method, the risk assessment of each failure mode is made by evaluating the parameters with respect to severity (S), occurrence (O), and Detection (D). RPN is obtained by multiplying these parameters. The higher the RPN value, the higher the risk is considered; thus, it should be considered risk mitigation. Although RPN is an effective way to assess risks in practice, this assessment has several drawbacks (Zandi et al. 2020; Wang et al. 2018). It has been criticized by many authors (Gul et al. 2020; Başhan et al. 2020; Mandal and Maiti 2014; Yang et al. 2008). Different combinations of different risk parameters can come together to reach the same RPN level (Liu et al. 2011; Boran and Gökler 2020). Prioritizing failure modes in FMEA with respect to RPN is a process that requires multi-criteria decision-making (MCDM) analysis (Braglia et al. 2003). MCDM is an advantageous approach that can structure the risk analysis process by separating it into stages and enumerate risk factors by considering their importance. Therefore, especially in recent years, MCDM methods have been used in FMEA to avoid the disadvantage of traditional RPN calculation (Liu et al. 2019). Many studies integrate MCDM with FMEA in order to avoid the limits of classical FMEA. A detailed review was presented by (Liu et al. 2019).

It is aimed to evaluate alternatives among many criteria in MCDM methods. The evaluation is made by one or more decision-makers (DM), and the preferences of DM are revealed. Alternatives are ranked, graded, or selected (Mohammadi and Rezaei 2020). In the literature, there are many methods such as analytic hierarchy process (AHP) (Ak and Gul 2019; Gul 2018), analytic network process (ANP) (Khan et al. 2020; Matin et al. 2020), multi-attribute rating technique (SMART) (Fitriani et al. 2020; Siregar et al. 2017) that determine the weight of decision criteria based on the preference of DM. A pairwise comparison-based MCDM, called BWM in recent years, was proposed by Rezaei (2015). BWM is becoming widespread day by day because it requires less data, can make more consistent comparisons, and gives more consistent results (Mi et al. 2019).

3.2 Literature Review

FMEA has taken its place in the literature as a systematic method widely used in analyzing the modes and effects of failures that occur in processes, systems, or product/service of a production/service system. There are some drawbacks in calculating the RPN, formulated as a combination of the three-parameter structure in the classical FMEA. Numerous studies have been proposed in the literature to overcome these drawbacks. New and original approaches that use multi-criteria decision analysis-based methods and their integration with the concepts such as fuzzy set theory, gray theory, soft set theory, and neutrosophic set theory have developed FMEA (Liu et al. 2019, 2013). The drawbacks of the RPN logic that exist in classical FMEA, revealed in the literature, can be listed as follows (Başhan et al. 2020; Qin et al. 2020; Bhattacharjee et al. 2020; Wang et al. 2020; Rezaee et al. 2020; Baykasoğlu and Gölcük 2020; Fattahi et al. 2020; Lo et al. 2020; Gul et al. 2020; Di Bona et al. 2018; Ozdemir et al. 2017; Liu et al. 2019, 2013; Bozdag et al. 2015; Park et al. 2018; Liu 2016):

- Apart from three parameters (S, O, and D), additional parameters that impact risk prioritization have not been fully considered (Liu et al. 2019; Di Bona et al. 2018). Therefore, parameters such as economic loss (e.g., percentage of the total annual budget fixed by the company for occupational health and safety measures), prevention, sensitivity to non-usage of personal protective equipment, sensitivity to non-implementation of reactive and proactive care, and the effectiveness of prevention measures and strategies must be functions of risk in an FMEA study (Seiti et al. 2020; Du et al. 2016; Lo et al. 2019).
- Weights of three parameters are not considered in RPN calculation in classical FMEA (Park et al. 2018; Liu et al. 2013; Huang et al. 2017). To overcome this drawback and provide a weighted assessment formula, some multi-criteria methods, including pairwise comparison, assess the decision criteria (e.g., AHP, BWM) and can be used.
- Different S, O, and D ratings may result in different meanings in the same RPN. However, risk priorities are definitely different (Huang et al. 2017; Catelani et al. 2018; Du et al. 2016; Safari et al. 2016).

- S, O, and D parameters are not easy to study precisely because of their subjective evaluation on a scale of 1–10. Using language terms in fuzzy numbers can better guide FMEA (Zhang et al. 2020; Mete 2019; Ozdemir et al. 2017; Zhao et al. 2017; Lo et al. 2019; Kutlu & Ekmekçioğlu 2012). More deficiencies can be found in Liu et al. (2013) and Liu et al. (2019). Both studies include two important literature reviews of FMEA-based studies.

In this study, to eliminate the drawbacks of the classic RPN calculation in FMEA, a modified approach is proposed. It is based on the BWM method, which is recently developed multi-criteria decision method by Rezaei (2015). Till its initial emergence, it has widely been applied to many areas, from production to the service industry (Mi et al. 2019). In this study, importance weights and preference values of failure modes with respect to these three parameters are determined via BWM. In the literature, many scholars have applied BWM in FMEA-based risk analysis problems. Table 3.1 shows an overview of the previous studies which apply BWM and its derivatives under the concept of FMEA. Among these, some studies (Dorosti et al. 2020; Cheng et al. 2020; Khalilzadeh et al. 2020; Nie et al. 2019, 2018; Ghoushchi et al. 2019; Chang et al. 2019; Tian et al. 2018; Peko et al. 2018; Lo and Liou 2018) use an auxiliary concept such as fuzzy MOORA, MULTIMOORA, Z-MOORA, GRA, VIKOR, TOPSIS, WASPAS, and COPRAS.

In our proposed approach, parameters of FMEA (S, O, and D) are assessed and then weighted by an expert. Hereafter, with respect to each risk parameter, six failure modes are assessed, and a preference vector for each assessment is gained. RPN values are finally calculated by multiplying the preference matrices and weight matrix. This systematic approach is tested in a plastic production facility, making its production via injection molding in Germany. Preventive measures are also in operation for failure modes considering their RPN values.

In the lights of abovementioned overview of previous studies integrating FMEA and BWM, the innovations may be shortened as follows:

(1) BWM is utilized to eliminate a drawback of FMEA related to the absence of weight values of risk parameters. At the same time, evaluations of each failure are realized with respect to each parameter by the aid of BWM computation logic to obtain preference values.
(2) Obtained preference and weight values from BWM are then merged to determine final RPN of each FM.
(3) The proposed risk assessment model brings a new framework for occupational risk assessment in manufacturing.
(4) This study identifies and discusses six potential failure modes in a plastic injection molding facility in Germany.
(5) The results of this study can support practitioners and risk analysts in formulating the improvement measures.
(6) The execution of the model comparison illustrates the advantages of the proposed model.

Table 3.1 Analysis of previous studies integrating FMEA and BWM

Study	Approach			Other concept	Novelty
	FMEA	BWM	Fuzzy extension?		
Dorosti et al. (2020)	√	√	Triangular fuzzy number (TFN)	Fuzzy MOORA	*Criterion weights determined by FBWM *FMs ranked by fuzzy MOORA
Gul et al. (2020)	√	√	TFN	–	*FMEA parameters determined hierarchically *Combined with the fuzzy rule-based and Bayesian network *An RPN calculation algorithm used
Cheng et al. (2020)	√	√	Trapezoidal neutrosophic number	MULTIMOORA	*Subjective and objective weights calculated by BWM *MULTIMOORA used for RPN calculation in surgical procedures
Khalilzadeh et al. (2020)	√	√	TFN	GRA, VIKOR	*FMEA integrated with GRA and VIKOR *Performance indicators for each risk-weighted by BWM *Strategies determined by a multi-objective mathematical program
Kolagar et al. (2020)	√	√	TFN	–	*Surgical cancelation factors weighted by FBWM *Computed a fuzzy RPN
Momen et al. (2019)	√	√	TFN	–	*FMs in hemodialysis weighted by FBWM

(continued)

Table 3.1 (continued)

Study	Approach			Other concept	Novelty
	FMEA	BWM	Fuzzy extension?		
Rastayesh et al. (2019)	√	√	–	–	*The BN constructed based on the FMEA criteria
Ghoushchi et al. (2019)	√	√	TFN	Z-MOORA	*A new FMEA approach introduced *The weighting of the risks performed by FBWM *Failures prioritized by a Z-MOORA
Chang et al. (2019)	√	√	TFN	TOPSIS, WASPAS	*FMEA based multi-attribute decision-making implemented *FMs weighted by FBWM *FMs ranked by TOPSIS *A WASPAS-based methodology proposed for RPN calculation
Tian et al. (2018)	√	√	TFN	VIKOR	*Risk factors weighted by FBWM *The rank of risks determined by fuzzy VIKOR
Peko et al. (2018)	√	√	–	TOPSIS	*Cost criterion attached to FMEA parameters *The parameters calculated by BWM *FMs ranked by TOPSIS
Liu et al. (2018)	√	√	–	–	*BWM under two-dimensional uncertain linguistic are used to weight risk factors

(continued)

Table 3.1 (continued)

Study	Approach			Other concept	Novelty
	FMEA	BWM	Fuzzy extension?		
Lo and Liou (2018)	√	√	TFN	GRA	*The cost variable added to the RPN calculation *The general RPN calculation modified *BWM and GRA integration used *FMEA carried out for electronics industry
Nie et al. (2018)	√	√	Fuzzy number	COPRAS	*BWM used to determine the weights *COPRAS is extended for ranking FMs

3.3 Methods

3.3.1 MCDM for Risk Assessment

This section introduces the importance of MCDM methods for the risk assessment problem and a flow of the process of injecting MCDM into classical risk analysis techniques. MCDM is an operations research concept that includes many methods for selecting the best alternative, prioritizing, and classifying alternatives as a result of a systematic and mathematical series of steps. As with other decision problems, MCDM is looking for solutions to many problems related to risk assessment and management. The decision-making procedure for risk assessment requires considering a range of hazards or types of hazards based on different risk parameters. For this purpose, MCDM methods have been suggested in recent years as a powerful tool to assist decision-makers in prioritizing risks and to reduce risks to an acceptable level MCDM-based risk analysis applications are increasing day by day. Risk assessment and management includes many elements with different goals and criteria. The main feature of MCDM methods is flexibility over the judgments of the decision-maker/makers. These methods aim to reach the ideal decision by assigning performance scores and weights. Figure 3.1 demonstrates the flow of the process of injecting MCDM into a usual risk assessment procedure. Here, "risk parameter" can refer to the elements of a classical risk analysis tool. As an example, in a Fine-Kinney procedure, these are probability, exposure, and consequence. In FMEA, severity, occurrence, and detection are the core parameters. Other components of this process include hazard list (with their associated risk descriptions), MCDM method for risk

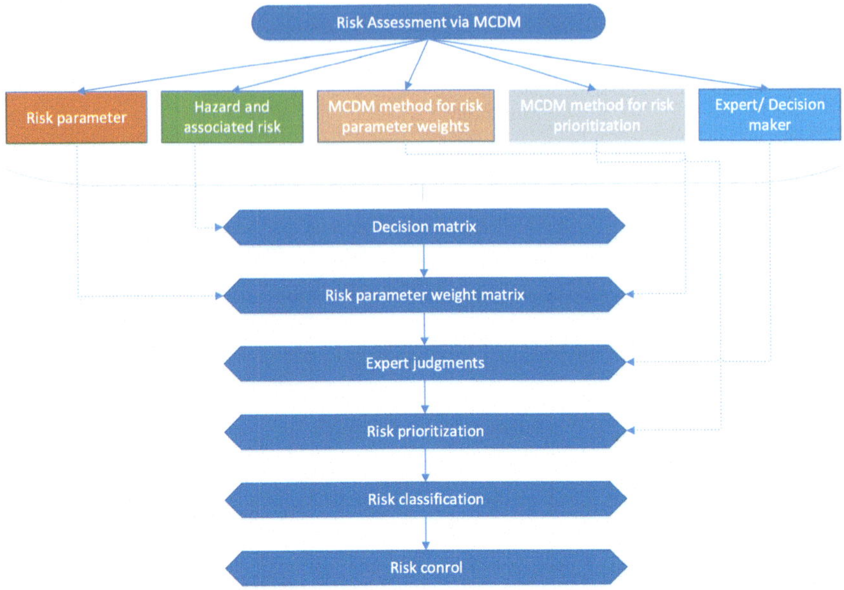

Fig. 3.1 Components of risk assessment via MCDM

parameter weighting (e.g., AHP, ANP, BWM, DEMATEL), MCDM method for risk prioritization (e.g., TOPSIS, VIKOR, WASPAS, GRA, COPRAS, MOORA), and decision-maker/expert.

3.3.2 FMEA

The FMEA is commonly implemented in the risk assessment process as a powerful method for risk assessment and reliability analysis that is developed for the aerospace industry in the 1960s at Grumman Aircraft Corporation (Bowles and Peláez 1995; Stamatis 2003). The preidentified failure modes' risk priority orders are determined by the RPN approach (Liu et al. 2019; Chin et al. 2009). A ten point-scale is used for each parameter in FMEA. Limitations to the conventional FMEA method lead to the emergence of some new FMEA-based approaches. While in some studies, MCDM is merged with FMEA, artificial intelligence, inference systems, soft computing, and some miscellaneous tools are also integrated with FMEA (Chai et al. 2016).

3.3.3 BWM

BWM was proposed by Rezaei (2015) to solve MCDM problems. BWM is a pairwise comparison-based weighting method. The proposed method is beneficial in a way. (*i*) Decision-makers determine the best and worst criteria among all criteria, pairwise comparison of best criterion with other criteria and worst criterion with others. There is no need for a pairwise comparison for all criteria. (*ii*) Some of the pairwise comparison-based MCDM methods use single vectors. (e.g., Swing and SMART family) Although these methods are data and time-efficient, they do not allow consistency checks. Some pairwise comparison-based methods (e.g., AHP) require a full pairwise comparison matrix. These methods allow consistency check, but they are not data and time-efficient. BWM requires less pairwise comparison compared to methods. It also allows for consistency check by having best to others and other to worst vectors. BWM is superior to other MCDMs in these aspects (Rezaei et al. 2016; Rezaei 2020).

Step 1. The criteria to be evaluated are determined. The criteria to be used in decision making are shown with $(c_1, c_2 \ldots, c_n)$.

Step 2. Best (most significant, most desired) and worst (least significant, least desired) criteria are determined among the determined criteria. Pairwise comparison is not performed at this stage.

Step 3. Using the numbers 1–9, it is determined how the best criterion differs from other criteria. The Best to other vector is created as:

$$A_B = (a_{B1}, a_{B2}, \ldots, a_{Bn})$$

where a_{Bj} shows the predilection of the best criterion B over criterion j Comparison of the criteria with themselves $(a_{BB} = 1)$

Step 4. Using the numbers 1–9, it is determined how the worst criterion differs from other criteria. Others-to-Worst vector is created as:

$$A_B = (a_{1W}, a_{2W}, \ldots, a_{nW})$$

where a_{jW} shows the predilection of the criterion j over the worst criterion W.

Step 5. Determination of weight $((W_1^*, W_2^*, \ldots W_n^*)$.

The optimum weight for the criteria is the one where for each pair of w_B/w_j and w_j/w_w we have $w_B/w_j = a_{jw}$. To satisfy these for all j, we should find a solution where the maximum absolute differences $\left|\frac{w_B}{w_j} - a_{Bj}\right|$ and $\left|\frac{w_j}{w_W} - a_{jW}\right|$ for all j is minimized. Given that the variables cannot be negative, and the sum of the variables is equal to one, the problem to be solved is:

$$\min_{j}\max\left\{\left|\frac{w_B}{w_j}-a_{Bj}\right|,\left|\frac{w_j}{w_W}-a_{jW}\right|\right\}$$

s.t

$$\sum w_j = 1$$

$w_j \geq 0$ for all j.

With the necessary conversion done, the problem is:

$$\min \xi$$

$\left|\frac{w_B}{w_j}-a_{Bj}\right| \leq \xi$ for all j.

$\left|\frac{w_j}{w_W}-a_{jW}\right| \leq \xi$ for all j

$$\sum w_j = 1$$

$w_j \geq 0$, for all j.

Solving problem, the optimum weights $((W_1^*, W_2^*, \ldots W_n^*)$ and ξ^* are calculated. Following the procedure in Rezaei (2015), the Consistency Ratio (CR) is calculated. The higher the ξ^*, higher CR and less reliable results will be obtained.

As a result of the solution of the problem, the variable weights $((W_1^*, W_2^*, \ldots W_n^*)$ and ξ^* are calculated. Then the consistency ratio is calculated. When the number of variables exceeds three, CR can never be equal to zero. It can be said that the lower the CR, the more consistent the evaluation is made.

3.3.4 Proposed Framework

The proposed FMEA framework is based on BWM method. The initial steps are about preparation for FMEA (determine failure modes and define RPN elements). The failure modes are identified that cause faulty products in the observed manufacturing plant. Then, the importance weights of the RPN elements and ranking of failure modes are calculated using BWM procedure. Preference values of each failure mode are computed with respect to S, O, and D. The flowchart of this proposed framework is provided in Fig. 3.2.

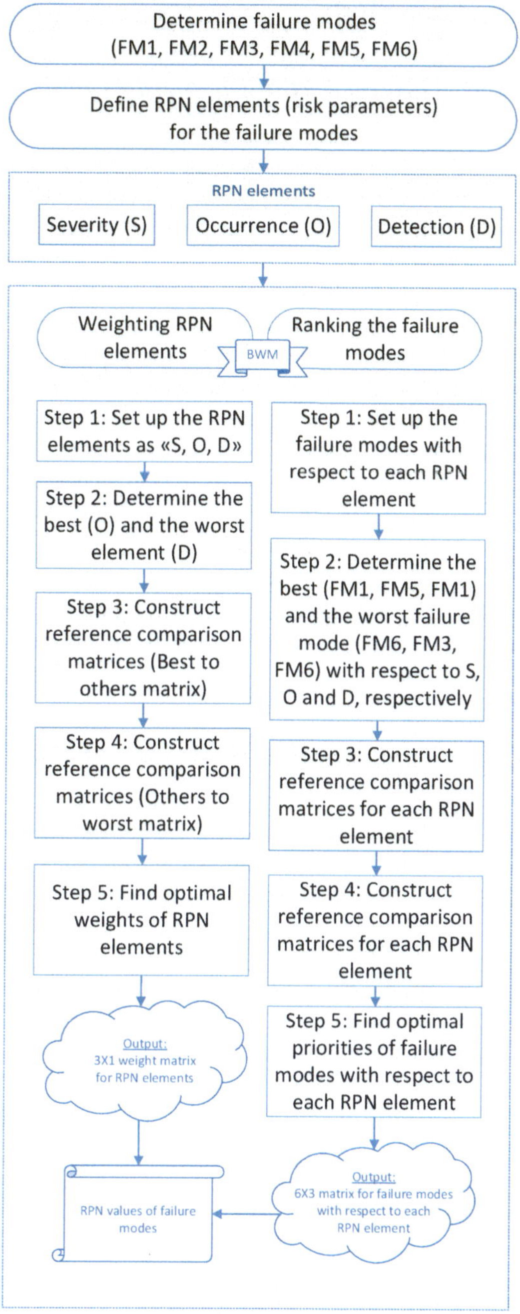

Fig. 3.2 The flowchart of the proposed framework

3.4 Case Study

3.4.1 Observed Manufacturing Plant

The observed manufacturing plant in which we performed the case study application is located in Germany. It makes plastic production by injection molding. The production includes several processes: Recycled plastic in crushed form is supplied from the supplier in the form of the bale. The bales are divided into pieces for homogenization. Thus, filling material can be added to the mixture. The raw material is brought to the shredder section to reduce the grain size. Raw material with reduced grain size is brought to the grinder. Large particles are removed from the system, and fine particles are brought to the centrifugal washing section. Here, at high temperature, solid plastic particles are melted to become liquid. Liquidized raw material is taken to the silo. Necessary additives are added to the molten raw material in the silo. With the help of the feeder, the raw material is brought to the extrusion section. In this section, the raw material is extracted in strip form. To prevent distortions in the product and achieve homogeneous cooling, the extruded strips are taken into the cooling pool. The extruded raw material is crushed in the grinder and brought to the silo. The final product is produced in the determined mold by sending the raw material with reduced particle size to the injection machine.

3.4.2 Application of the Proposed Framework

The implementation of BWM in this FMEA study concerns with assigning importance weights to the RPN elements and ranking of failure modes. Each failure mode's preference values are computed with respect to RPN elements, which are S, O, and D.

3.4.2.1 Determination of Importance Weights of RPN Elements

In this phase of the proposed framework, expert assessment has been made to determine importance weights for S, O, and D. BWM has been used to make calculations. As an example, the evaluation regarding the reference comparison between RPN elements is presented in Fig. 3.3.

By following the BWM section's steps, the weight values for each RPN element are computed as in Fig. 3.4.

3.4.2.2 Ranking of Failure Modes

Similarly, six failure modes were evaluated with respect to RPN elements. They are explained in detail as follows:

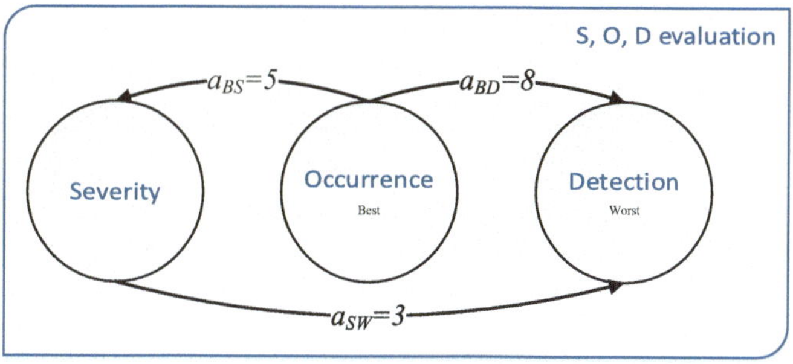

Fig. 3.3 Reference comparison in assessing RPN elements by BWM

Fig. 3.4 Importance weights of RPN elements by BWM implementation

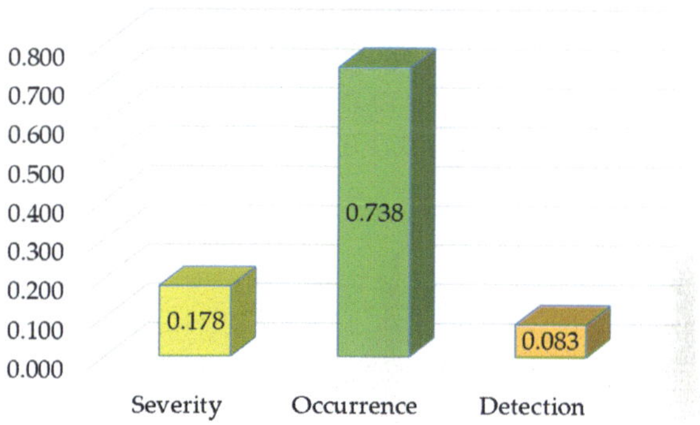

FM1-Failure to send the appropriate quantity or type of products to the Shredder: In this case, the shredder ampere rises, and the shredder may stop because the density of the raw material sent can be high.

FM2-Blockage of the pipe at the grinder inlet: It causes the temperature of the raw material to rise. The raw material that goes to the crushing section at high temperature is compressed in this section.

FM3-Raw material compression in centrifuges: Combustion and bad odor are emitted in the centrifuges due to the compression of the raw material. The burnt raw material must be removed from the system.

FM4-Clogging of connection pipes of silos: The connection pipes are clogged due to the accumulation of excess raw material in the silos. If preventive measures are not taken, the raw material prevents the augers at the outlet of the centrifuge to function.

FM5-Extruder blocks the flow of raw materials: Since the extruder head is not cleaned well or the raw material is sent at high temperature, the raw material adheres to each other.

FM 6-Deterioration of press molds: Preventing the movement of the press molds when going up and down due to expansion and contraction if the raw material sent to the presses is more than 220 °C.

The evaluations with respect to S, O, and D elements are presented in Figs. 3.5, 3.6, and 3.7, respectively. By following the steps as in the previous sub-section, the preference values of six failure modes according to three RPN elements are presented in Fig. 3.8.

Two matrices are obtained at the end of these two implementation processes in the BWM. The first matrix consists of three columns. These columns consist of preference values of failure modes according to RPN elements, as presented in Table 3.2. The second matrix includes the importance weights of RPN elements, as given in Fig. 3.4. By multiplying these two matrices, RPN values of failure modes are obtained as follows:

$$\begin{bmatrix} 0.336 & 0.311 & 0.358 \\ 0.262 & 0.097 & 0.277 \\ 0.084 & 0.045 & 0.167 \\ 0.168 & 0.078 & 0.087 \\ 0.112 & 0.402 & 0.070 \\ 0.037 & 0.066 & 0.040 \end{bmatrix} \times \begin{bmatrix} 0.178 \\ 0.73 \\ 0.083 \end{bmatrix} = \begin{bmatrix} 0.320 \\ 0.142 \\ 0.062 \\ 0.095 \\ 0.323 \\ 0.058 \end{bmatrix}$$

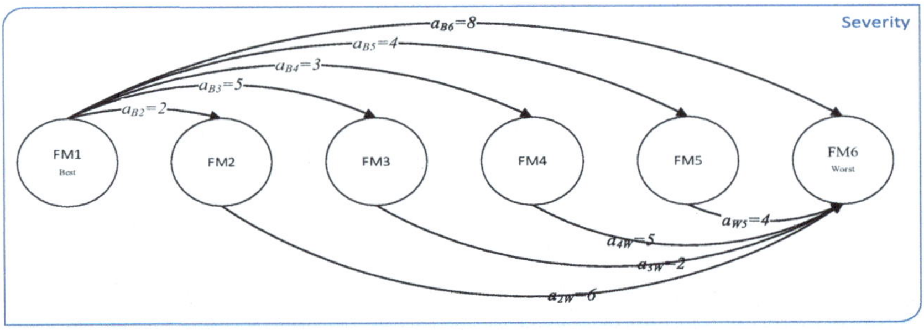

Fig. 3.5 Evaluation of failure modes with respect to "severity" element

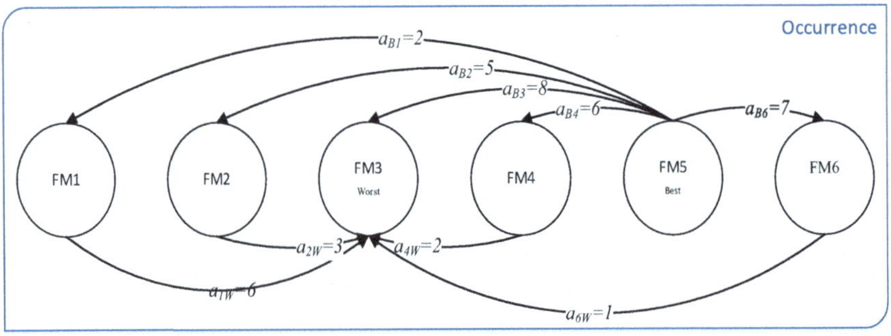

Fig. 3.6 Evaluation of failure modes with respect to "occurrence" element

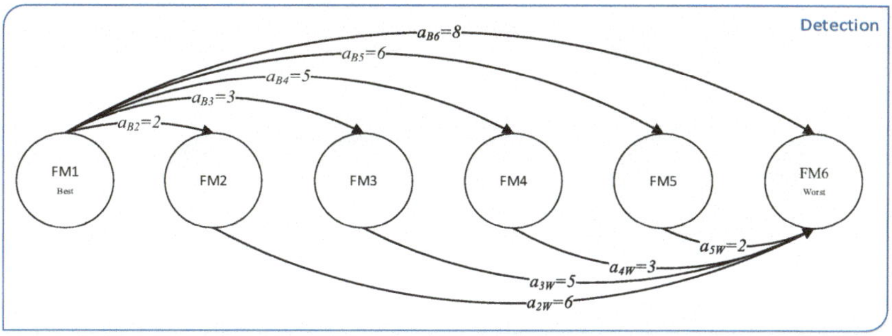

Fig. 3.7 Evaluation of failure modes with respect to "detection" element

The consistency check is made for each matrix in the proposed framework. All four matrices have a CR value of lower than 0.22 (Table 3.3). It is acceptable for BWM.

3.4.3 Analysis of the Result

The failure prioritization of six failure modes is $FM5 > FM1 > FM2 > FM4 > FM3 > FM6$. The failure mode of FM5 (Extruder blocking the flow of raw materials) with its highest final RPN score ($RPN = 0.323$) should be taken more attention. On the other hand, FM1 is also determined as the second most important failure mode ($RPN = 0.320$). FM3 and FM6 are determined as the least important failure modes.

The main factor in the formation of FM5 is that the raw material taken from recycling is not homogeneous. It is quite challenging to adjust the melt's optimum temperature due to plastics with different melting temperatures in the supplied bale. In this case, the following measures can be taken. (*i*) As much as possible, less molten plastic should be sent to the system. This will reduce the production rate but improve flow through the extruder. (*ii*) It is not possible to provide the production parameters because the raw material used is not homogeneous. Therefore, preliminary tests should be carried out at each supply to determine the optimum melting temperature and pressure amount. Thus, congestion in the extruder is prevented. (*iii*) Homogeneity should be taken into account in supplier selection. (*iv*) The grain size in the grinder, feeder speed, silo temperature should be controlled continuously, and necessary precautions should be taken to prevent clogging of the extruder.

The main factor in the occurrence of FM1 is the size difference in the raw material supplied. To prevent this failure: (*i*) A pre-screening process should be carried out on the raw material supplied. Thus, more homogeneous particles will not create clogging in the shredder. (*ii*) Storing the raw material at appropriate humidity and temperature will increase the shredder performance. (*iii*) Necessary thread adjustments should

	Severity	Occurrence	Detection
Severity	1	0.2	3
Occurrence	5	1	8
Detection	0.333	0.125	1

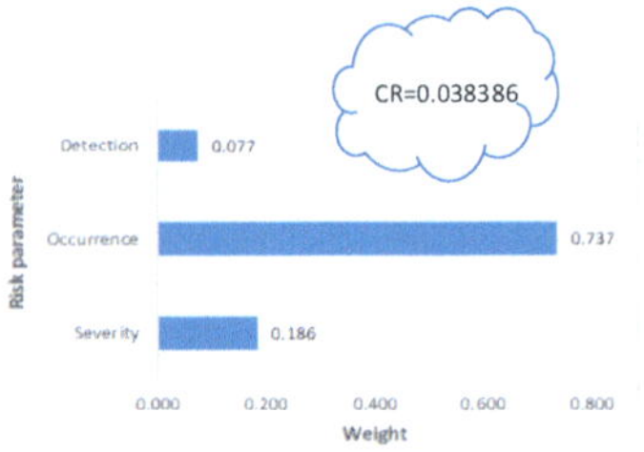

Evaluations of FMs with respect to Severity	FM1	FM2	FM3	FM4	FM5	FM6
FM1	1	2	5	3	4	8
FM2	0.5	1	3	2	2	6
FM3	0.2	0.333	1	2	2	1
FM4	0.333	0.5	0.5	1	0.5	1
FM5	0.25	0.5	0.5	2	1	0.5
FM6	0.125	0.167	1	1	2	1

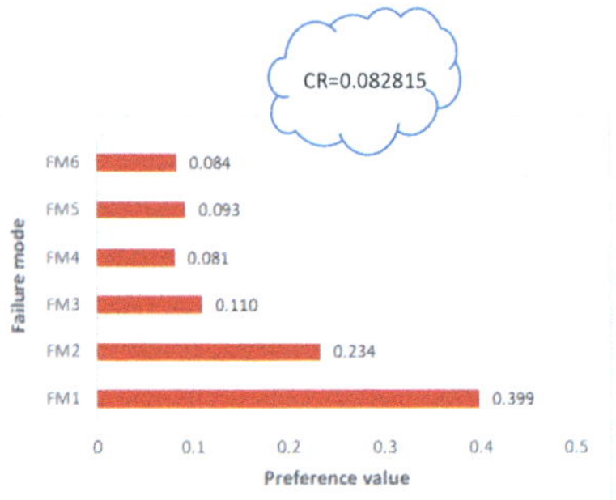

Evaluations of FMs with respect to Occurrence	FM1	FM2	FM3	FM4	FM5	FM6
FM1	1	3	6	4	0.2	5
FM2	0.333	1	3	2	0.5	2
FM3	0.167	0.333	1	0.333	0.125	0.5
FM4	0.25	0.5	3	1	0.167	2
FM5	5	2	8	6	1	7
FM6	0.2	0.5	2	0.5	0.143	1

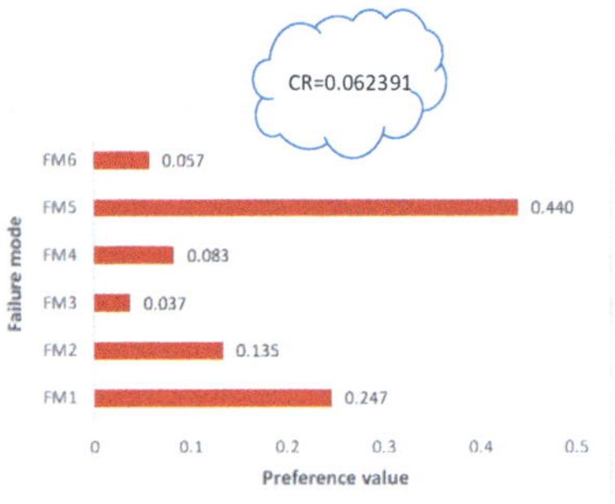

Evaluations of FMs with respect to Detection	FM1	FM2	FM3	FM4	FM5	FM6
FM1	1	2	3	5	6	8
FM2	0.5	1	2	3	4	6
FM3	0.333	0.5	1	2	3	4
FM4	0.2	0.333	0.5	1	2	3
FM5	0.167	0.25	0.333	0.5	1	2
FM6	0.125	0.167	0.2	0.333	0.5	1

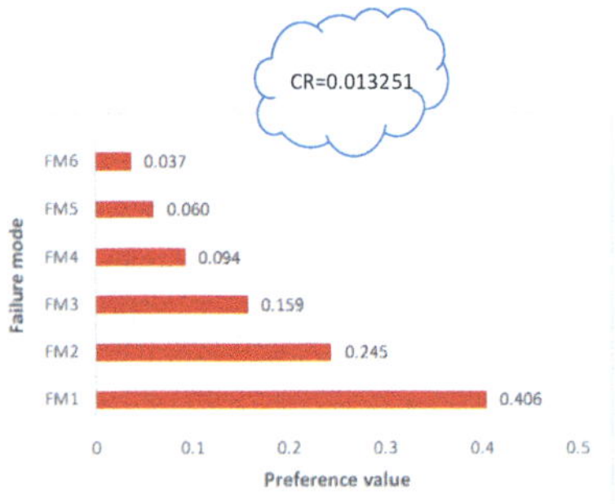

Fig. 3.8 The evaluations and obtained weight and preference values (including CR results)

Table 3.2 The preference values of six failure modes according to RPN elements

Failure mode	Severity	Occurrence	Detection
FM1	0.336	0.311	0.358
FM2	0.262	0.097	0.277
FM3	0.084	0.045	0.167
FM4	0.168	0.078	0.087
FM5	0.112	0.402	0.070
FM6	0.037	0.066	0.040

Table 3.3 The consistency ratio of each matrix

Matrix	ξ	CR
RPN element weight matrix	0.859	0.19
Failure mode matrix with respect to severity	0.999	0.22
Failure mode matrix with respect to occurrence	0.859	0.19
Failure mode matrix with respect to detection	0.859	0.19

be made according to the incoming raw material while setting the shredder. The operators should be given the necessary training in this regard. Since the raw material arrival is irregular, the operator should make the necessary adjustment without delay.

The emergence of FM2 is closely related to FM1. The measures to be taken for FM1 will reduce the occurrence of FM2 as well. Besides, adjusting the grinder tooth size and speed appropriately will reduce their failure in this section.

The occurrence of FM4 is usually caused by operators. Therefore, this failure is to be minimized. (*i*) Operators should be informed about periodic maintenance. (*ii*) While the system is shutting down, it should be run idle for a while, and it should be ensured that all molten plastic is separated from the pipes.

The occurrence of FM3 is directly related to FM1 and FM2. The measures to be taken to prevent FM1 and FM2 will reduce occurrence of FM3. Besides, measures similar to FM5 should be taken for temperature adjustments in this section.

To minimize FM6, temperature at which the melted plastic is sent to the press should be prevented from being higher than 220 °C. For this reason, temperature analysis should be done in the system, and if necessary, the melted plastic should be kept in the feeder for a time to cool.

3.5 Discussion

A comparative study that considers evaluations of both risk parameters and FMs with respect to three RPN elements via AHP is demonstrated to test the validity of the proposed approach. Figure 3.8 shows the whole evaluation and obtained weight

Table 3.4 RPN values of failure modes obtained from AHP

Failure mode	RPN	Rank
FM1	0.287	2
FM2	0.162	3
FM3	0.060	6
FM4	0.084	4
FM5	0.346	1
FM6	0.061	5

values and preference values of six FMs according to these parameters (S, O, and D). All evaluation matrices are found consistent. The CR values of each matrix are also given in Fig. 3.8. By combining preference values of failure modes with respect to each RPN element, the first matrix is gained. Then, final RPN values were obtained by multiplying this matrix and weight matrix. The results are presented in Table 3.4.

As a result of the AHP calculation, FM ranks are as follows: FM5 > FM1 > FM2 > FM4 > FM3 > FM6 When the values of this RPN calculation procedure by AHP and the proposed approach by BWM are compared, it is observed that the ranks are quite similar. Pearson correlation coefficient regarding final RPN values and Spearman rank correlation coefficient regarding rankings in both approaches were determined as 0.98 and 0.94, respectively. In this case, it can be said that there is a very strong relationship between these results.

3.6 Conclusion

Machine failures and production errors directly affect production performance. These cause production stoppages and increase in unexpected costs of the business. In recent years, various methods have been presented to reduce the effects of these uncertainties. FMEA is one of these methods. It is widely applied in the automotive, aerospace, rail, and nuclear industries due to its easy use and effective results. Risk parameters of severity, occurrence, and detection are multiplied to obtain RPN. The higher the RPN value, the higher the risk is considered; therefore, it should be considered risk reduction. Different MCDM methods are used to prioritize failure modes in FMEA in terms of the RPN.

In this study, we used BWM to prioritize the failure modes with the determined importance weights of the S, O, and D parameters. In our proposed approach, FMEA (S, O, and D) parameters are evaluated and then weighted by an expert. After that, six failure modes for each FMEA parameter are evaluated. A preference vector is obtained for each evaluation. In the final step of the proposed approach, RPN values are calculated by multiplying the preference matrices and the weight matrix. The proposed approach is analyzed in a plastic production facility in Germany that makes production by injection molding. Preventive measures are also in operation for failure modes considering their RPN values. The plastics industry is a competitive production area where unit production costs are a significant indicator. Faulty production causes financial losses for businesses but also harms the environment. Therefore, minimizing failures in this sector will benefit society in many ways.

In future studies, while calculating the weights of failure modes using different MCDM methods, authors plan to apply similar procedures for different raw materials and molds. Thus, it will be easier to plan preventive activities to be implemented by the enterprise to prevent failures in different types of products.

References

Ak MF, Gul M (2019) AHP–TOPSIS integration extended with Pythagorean fuzzy sets for information security risk analysis. Complex Intell Syst 5(2):113–126

Başhan V, Demirel H, Gul M (2020) An FMEA-based TOPSIS approach under single valued neutrosophic sets for maritime risk evaluation: the case of ship navigation safety. Soft Comput: 1–16

Baykasoğlu A, Gölcük İ (2020) Comprehensive fuzzy FMEA model: a case study of ERP implementation risks. Oper Res Int J 20(2):795–826

Bhattacharjee P, Dey V, Mandal UK (2020) Risk assessment by failure mode and effects analysis (FMEA) using an interval number based logistic regression model. Saf Sci 132: 104967

Boran S, Gökler SH (2020) A novel FMEA model using hybrid ANFIS–Taguchi method. Arab J Sci Eng 45(3):2131–2144

Bowles JB, Peláez CE (1995) Fuzzy logic prioritization of failures in a system failure mode, effects and criticality analysis. Reliab Eng Syst Saf 50(2):203–213

Bozdag E, Asan U, Soyer A, Serdarasan S (2015) Risk prioritization in failure mode and effects analysis using interval type-2 fuzzy sets. Expert Syst Appl 42(8):4000–4015

Braglia M, Frosolini M, Montanari R (2003) Fuzzy TOPSIS approach for failure mode, effects and criticality analysis. Qual Reliab Eng Int 19(5):425–443

Catelani M, Ciani L, Venzi M (2018) Failure modes, mechanisms and effect analysis on temperature redundant sensor stage. Reliab Eng Syst Saf 180:425–433

Chai KC, Jong CH, Tay KM, Lim CP (2016) A perceptual computing-based method to prioritize failure modes in failure mode and effect analysis and its application to edible bird nest farming. Appl Soft Comput 49:734–747

Chang TW, Lo HW, Chen KY, Liou JJ (2019) A novel FMEA model based on rough BWM and rough TOPSIS-AL for risk assessment. Mathematics 7(10):874

Cheng PF, Li DP, He JQ, Zhou XH, Wang JQ, Zhang HY (2020) Evaluating surgical risk using FMEA and MULTIMOORA methods under a single-valued trapezoidal neutrosophic environment. Risk Manage Healthcare Policy 13:865

Cheung WM, Leong JT, Vichare P (2017) Incorporating lean thinking and life cycle assessment to reduce environmental impacts of plastic injection moulded products. J Clean Prod 167:759–775

Chin KS, Wang YM, Poon GKK, Yang JB (2009) Failure mode and effects analysis by data envelopment analysis. Decis Support Syst 48(1):246–256

Di Bona G, Silvestri A, Forcina A, Petrillo A (2018) Total efficient risk priority number (TERPN): a new method for risk assessment. J Risk Res 21(11):1384–1408

Dorosti S, Fathi M, Ghoushchi SJ, Khakifirooz M, Khazaeili M (2020) Patient waiting time management through fuzzy based failure mode and effect analysis. J Intell Fuzzy Syst (Preprint): 1–12

Du Y, Lu X, Su X, Hu Y, Deng Y (2016) New failure mode and effects analysis: an evidential downscaling method. Qual Reliab Eng Int 32(2):737–746

Fattahi R, Tavakkoli-Moghaddam R, Khalilzadeh M, Shahsavari-Pour N, Soltani R (2020) A novel FMEA model based on fuzzy multiple-criteria decision-making methods for risk assessment. J Enterp Inf Manage

Fitriani N, Suzanti IO, Jauhari A, Khozaimi A (2020) Application monitoring and evaluation using SMART (Simple Multi attribute Rating Technique) method. In: Journal of Physics: Conference Series, vol 1569, no 2, p 022090. IOP Publishing

Fuentes-Huerta MA, González-González DS, Cantú-Sifuentes M, Praga-Alejo RJ (2018) RCM implementation on plastic injection molding machine considering correlated failure modes and small size sample. Int J Adv Manuf Technol 95(9–12):3465–3473

Ghoushchi SJ, Yousefi S, Khazaeili M (2019) An extended FMEA approach based on the Z-MOORA and fuzzy BWM for prioritization of failures. Appl Soft Comput 81: 105505

Gökler SH, Boran S (2020) An integrated SMED-fuzzy FMEA model for reducing setup time. J Intell Manuf: 1–15

Gul M (2018) Application of Pythagorean fuzzy AHP and VIKOR methods in occupational health and safety risk assessment: the case of a gun and rifle barrel external surface oxidation and colouring unit. Int J Occup Saf Ergon: 1–14

Gul M, Yucesan M, Celik E (2020) A manufacturing failure mode and effect analysis based on fuzzy and probabilistic risk analysis. Appl Soft Comput 96: 106689

Huang J, Li ZS, Liu HC (2017) New approach for failure mode and effect analysis using linguistic distribution assessments and TODIM method. Reliab Eng Syst Saf 167:302–309

Kang K, Subramaniam V (2018) Integrated control policy of production and preventive maintenance for a deteriorating manufacturing system. Comput Ind Eng 118:266–277

Karasu MK, Salum L (2018) FIS-SMED: a fuzzy inference system application for plastic injection mold changeover. Int J Adv Manuf Technol 94(1–4):545–559

Khalilzadeh M, Katoueizadeh L, Zavadskas EK (2020) Risk identification and prioritization in banking projects of payment service provider companies: an empirical study. Front Bus Res China 14(1):1–27

Khan SA, Alenezi M, Agrawal A, Kumar R, Khan RA (2020) Evaluating performance of software durability through an integrated fuzzy-based symmetrical method of ANP and TOPSIS. Symmetry 12(4):493

Kolagar M, Hosseini SMH, Felegari R (2020) Developing a new BWM-based GMAFMA approach for evaluation of potential risks and failure modes in production processes. Int J Qual Reliab Manage

Kutlu AC, Ekmekçioğlu M (2012) Fuzzy failure modes and effects analysis by using fuzzy TOPSIS-based fuzzy AHP. Expert Syst Appl 39(1):61–67

Li GF, Li Y, Chen CH, He JL, Hou TW, Chen JH (2020) Advanced FMEA method based on interval 2-tuple linguistic variables and TOPSIS. Qual Eng 32(4):653–662

Liu HC (2016) FMEA using uncertainty theories and MCDM methods. In: FMEA using uncertainty theories and MCDM methods. Springer, Singapore, pp 13–27

Liu HC, Chen XQ, Duan CY, Wang YM (2019) Failure mode and effect analysis using multi-criteria decision making methods: a systematic literature review. Comput Ind Eng 135:881–897

Liu HC, Liu L, Bian QH, Lin QL, Dong N, Xu PC (2011) Failure mode and effects analysis using fuzzy evidential reasoning approach and grey theory. Expert Syst Appl 38(4):4403–4415

Liu HC, Liu L, Liu N (2013) Risk evaluation approaches in failure mode and effects analysis: a literature review. Expert Syst Appl 40(2):828–838

Liu Y, Li F, Wang Y, Yu X, Yuan J, Wang Y (2018) Assessing the environmental impact caused by power grid projects in high altitude areas based on BWM and Vague sets techniques. Sustainability 10(6):1768

Lo HW, Liou JJ (2018) A novel multiple-criteria decision-making-based FMEA model for risk assessment. Appl Soft Comput 73:684–696

Lo HW, Liou JJ, Huang CN, Chuang YC (2019) A novel failure mode and effect analysis model for machine tool risk analysis. Reliab Eng Syst Saf 183:173–183

Lo HW, Shiue W, Liou JJ, Tzeng GH (2020) A hybrid MCDM-based FMEA model for identification of critical failure modes in manufacturing. Soft Comput

Mandal S, Maiti J (2014) Risk analysis using FMEA: fuzzy similarity value and possibility theory based approach. Expert Syst Appl 41(7):3527–3537

Matin A, Zare S, Ghotbi-Ravandi M, Jahani Y (2020) Prioritizing and weighting determinants of workers' heat stress control using an analytical network process (ANP) a field study. Urban Clim 31: 100587

Mete S (2019) Assessing occupational risks in pipeline construction using FMEA-based AHP-MOORA integrated approach under Pythagorean fuzzy environment. Hum Ecol Risk Assess Int 25(7):1645–1660

Mi X, Tang M, Liao H, Shen W, Lev B (2019) The state-of-the-art survey on integrations and applications of the best worst method in decision making: why, what, what for and what's next? Omega 87:205–225

Mohammadi M, Rezaei J (2020) Bayesian best-worst method: a probabilistic group decision making model. Omega 96: 102075

Momen S, Tavakkoli-Moghaddam R, Ghasemkhani A, Shahnejat-Bushehri S, Tavakkoli-Moghaddam H (2019) Prioritizing surgical cancellation factors based on a fuzzy best-worst method: a case study. IFAC-PapersOnLine 52(13):112–117

Nie RX, Tian ZP, Wang XK, Wang JQ, Wang TL (2018) Risk evaluation by FMEA of supercritical water gasification system using multi-granular linguistic distribution assessment. Knowl-Based Syst 162:185–201

Nie W, Liu W, Wu Z, Chen B, Wu L (2019) Failure mode and effects analysis by integrating Bayesian fuzzy assessment number and extended gray relational analysis-technique for order preference by similarity to ideal solution method. Qual Reliab Eng Int 35(6):1676–1697

Ozdemir Y, Gul M, Celik E (2017) Assessment of occupational hazards and associated risks in fuzzy environment: a case study of a university chemical laboratory. Hum Ecol Risk Assess Int J 23(4):895–924

Pan E, Liao W, Xi L (2010) Single-machine-based production scheduling model integrated preventive maintenance planning. Int J Adv Manuf Technol 50(1–4):365–375

Park HS, Dang XP (2017) Development of a smart plastic injection mold with conformal cooling channels. Procedia Manuf 10:48–59

Park J, Park C, Ahn S (2018) Assessment of structural risks using the fuzzy weighted Euclidean FMEA and block diagram analysis. Int J Adv Manuf Technol 99(9–12):2071–2080

Peko M, Komatina N, Banduka N, Crnjac M (2018) The assessment and ranking of failures in the information technology industry based on FMEA and MCDM. Ekonomski Horizonti 20(3):257–268

Qin J, Xi Y, Pedrycz W (2020) Failure mode and effects analysis (FMEA) for risk assessment based on interval type-2 fuzzy evidential reasoning method. Appl Soft Comput 89: 106134

Rastayesh S, Bahrebar S, Blaabjerg F, Zhou D, Wang H, Dalsgaard Sørensen J (2019) A system engineering approach using FMEA and Bayesian network for risk analysis—a case study. Sustainability 12(1): 77

Rezaee MJ, Yousefi S, Eshkevari M, Valipour M, Saberi M (2020) Risk analysis of health, safety and environment in chemical industry integrating linguistic FMEA, fuzzy inference system and fuzzy DEA. Stoch Env Res Risk Assess 34(1):201–218

Rezaei J (2015) Best-worst multi-criteria decision-making method. Omega 53:49–57

Rezaei J (2020) A concentration ratio for nonlinear best worst method. Int J Inf Technol Decis Making: 1–17

Rezaei J, Nispeling T, Sarkis J, Tavasszy L (2016) A supplier selection life cycle approach integrating traditional and environmental criteria using the best worst method. J Clean Prod 135:577–588

Sadeghi BHM (2000) A BP-neural network predictor model for plastic injection molding process. J Mater Process Technol 103(3):411–416

Safari H, Faraji Z, Majidian S (2016) Identifying and evaluating enterprise architecture risks using FMEA and fuzzy VIKOR. J Intell Manuf 27(2):475–486

Seiti H, Fathi M, Hafezalkotob A, Herrera-Viedma E, Hameed IA (2020) Developing the modified R-numbers for risk-based fuzzy information fusion and its application to failure modes, effects, and system resilience analysis (FMESRA). ISA Trans
Siregar D, Arisandi D, Usman A, Irwan D, Rahim R.(2017) Research of simple multi-attribute rating technique for decision support. In: Journal of Physics: Conference Series, vol 930, no 1, p 012015. IOP Publishing
Stamatis DH (2003) Failure mode and effect analysis: FMEA from theory to execution. ASQ Quality Press, New York
Tian ZP, Wang JQ, Zhang HY (2018) An integrated approach for failure mode and effects analysis based on fuzzy best-worst, relative entropy, and VIKOR methods. Appl Soft Comput 72:636–646
Wang L, Yan F, Wang F, Li Z (2020) FMEA-CM based quantitative risk assessment for process industries—a case study of coal-to-methanol plant in China. Process Saf Environ Prot
Wang W, Liu X, Qin Y, Fu Y (2018) A risk evaluation and prioritization method for FMEA with prospect theory and Choquet integral. Saf Sci 110:152–163
Yang Z, Bonsall S, Wang J (2008) Fuzzy rule-based Bayesian reasoning approach for prioritization of failures in FMEA. IEEE Trans Reliab 57(3):517–528
Zandi P, Rahmani M, Khanian M, Mosavi A (2020) Agricultural risk management using fuzzy TOPSIS analytical hierarchy process (AHP) and failure mode and effects analysis (FMEA). Agriculture 10(11):504
Zhang Y, Wang Q, Liaw B, Nagpure SC, Dufek EJ, Dickerson CC (2020) A quantitative failure analysis on capacity fade in rechargeable lithium metal cells. J Electrochem Soc 167(9): 090502
Zhao H, You JX, Liu HC (2017) Failure mode and effect analysis using MULTIMOORA method with continuous weighted entropy under interval-valued intuitionistic fuzzy environment. Soft Comput 21(18):5355–5367
Zhou Q, Thai VV (2016) Fuzzy and grey theories in failure mode and effect analysis for tanker equipment failure prediction. Saf Sci 83:74–79

Chapter 4
Fuzzy Risk Assessment in the Presence of Uncertainties in Heterogeneous Preferences Elicitation and Reliability

Ahmad Syafadhli Abu Bakar, Mohd Zahurin Mohamed Kamali, Ku Muhammad Naim Ku Khalif, Alexander Gegov, and Siti Nazifah Zainol Abidin

Abstract Fuzzy risk assessment has displayed successful track records in mitigating the levels of risks in the presence of uncertainty. With incorporation of Z-numbers, the capability of fuzzy risk assessment methods in dealing with risks is enhanced as the element of reliability is contemplated. Although, established fuzzy risk assessment methods based on Z-numbers possess great quality in acknowledging the presence of uncertainty on the risk faced, the respects given are still not holistic. This is because they restrict the presence of uncertainty only to when the preferences elicited by the risk analysts are partially known. Nonetheless, the presence of uncertainty can also be found when preferences elicited by the risk analysts are partially unknown, completely known and completely unknown. For this reason, the capability of established fuzzy risk assessment methods based on Z-numbers is considered as partially complete as they ignore the existence of the heterogeneous nature of the preferences elicited by the risk analysts. Due to this, this chapter presents a novel fuzzy risk assessment method based on Z-numbers with incorporation of grey numbers

A. S. A. Bakar (✉) · M. Z. M. Kamali
Mathematics Division, Centre for Foundation Studies in Science, University of Malaya, Kuala Lumpur, Malaysia
e-mail: ahmadsyafadhli@um.edu.my

M. Z. M. Kamali
e-mail: mzmk@um.edu.my

K. M. N. K. Khalif
Centre for Mathematical Sciences, College of Computing and Applied Sciences, Universiti Malaysia Pahang, Pahang, Malaysia
e-mail: kunaim@ump.edu.my

A. Gegov
School of Computing, University of Portsmouth, Portsmouth, UK
e-mail: alexander.gegov@port.ac.uk

S. N. Z. Abidin
Faculty of Computer and Mathematical Sciences, Universiti Teknologi Mara, Jasin, Melaka, Malaysia
e-mail: sitinazifah@uitm.edu.my

J. Ren (ed.), *Multi-Criteria Decision Analysis for Risk Assessment and Management*, Industrial Ecology and Environmental Management 1,
https://doi.org/10.1007/978-3-030-78152-1_4

to holistically acknowledge the presence of uncertainty in the heterogeneous nature of preferences elicited by the risk analysts. In the proposed method, a novel fuzzy agreement relation approach is established to handle the interactions between the common and the uncommon heterogeneous preferences elicited by the risk analysts. Then, results obtained from this step are aggregated using a novel fuzzy risk evaluation rating method to assess the correct level of risks, such that the assessments are consistent with the presence of uncertainty in the heterogeneous preferences elicited by the risk analysts. Later on, theoretical validation and application on real world risk assessment problem in electrical arc welding industry are demonstrated to signify the novelty, validity and feasibility of the proposed work.

Keywords Fuzzy risk assessment · Z-numbers · Grey numbers · Reliability degree · Uncertain heterogeneous risk analyst's preferences elicitation

4.1 Introduction

Fuzzy risk assessment methods are developed to handle risks that involve uncertainty. One of the established fuzzy concepts that concerns with the uncertainty is the Z-numbers (Bakar and Gegov 2015; Kang et al. 2012; Zadeh 2011; Allahviranloo and Ezadi 2019). Established fuzzy risk assessment methods based on Z-numbers often describe each risk under consideration as an ordered pair of restriction of the preferences elicited by the risk analyst's and the reliability of the restriction (Jiang et al. 2017; Wu et al. 2018; Abiyev et al. 2018). In the literature on fuzzy risk assessment, incorporation of the pair (restriction and reliability components) has complemented established fuzzy risk assessment methods to successfully resolve numerous risk assessment problems such as risk assessment evaluations in failure mode of rotor blades of an aircraft turbine (Jiang et al. 2017), investigation on risk components in manufacturing and medical industries (Wu et al. 2018) and assessment of risk in food security (Abiyev et al. 2018). In order to define risks, established fuzzy risk assessment methods based on Z-numbers evaluate each risk under consideration based on two common risk factors, namely, the risk severity of loss and the risk probability of failure (Bakar et al. 2020; Zhao et al. 2020; Natha Reddy and Gokulachandran 2020; Chukwuma et al. 2020). Each of these factors that are usually expressed based on the preferences elicited by the risk analysts, is in this case represented by their own ordered pair of restriction and reliability (Jiang et al. 2017; Wu et al. 2018; Abiyev et al. 2018).

The literature on established fuzzy risk assessment methods based on Z-numbers signify that they possess great capability to deal with the presence of uncertainty (Marhamati et al. 2018; Peng et al. 2019; Hendiani et al. 2020; Azadeh and Kokabi 2016). However, their acknowledgement in terms of the presence of uncertainty on each risk under consideration is graded as partially complete. This is because established fuzzy risk assessment methods based on Z-numbers take into account only the presence of uncertainty when preferences elicited by the risk analysts are partially known (Jiang et al. 2017; Wu et al. 2018; Abiyev et al. 2018). Nonetheless, the presence of uncertainty can also happen when preferences elicited by the risk analysts are completely known, completely unknown and partially unknown (Bakar et al. 2020; Yang and John 2012; Huang et al. 2008). This points out that established fuzzy risk assessment methods based on Z-numbers do not have the holistic feature as they restrict the presence of uncertainty in the preferences elicited by the risk analysts to be homogeneous (partially known only), even if the presence of uncertainty is actually heterogeneous in nature (Bakar et al. 2020). Apart from that, the interactions between the common and uncommon heterogeneous preferences elicited by the risk analysts also indicate that the established fuzzy risk assessment methods based on Z-numbers are unable to holistically track the performance of risks in the presence of uncertainty. The above-mentioned inefficiencies of the established fuzzy risk assessment methods based on Z-numbers point out the motivations for this study.

In order to resolve the mentioned limitations of established fuzzy risk assessment methods based on Z-numbers, this study utlises the concept of grey number (Yang and John 2012; Bakar et al. 2019) to acknowledge the presence of uncertainty when preference elicited by the risk analysts are heterogeneous. This concept is suitable for this study because it defines the completely known preferences elicitation in the form of white number, partially known and unknown preferences elicitation as the grey number and completely unknown preferences elicitation as the black number (Baker et al. 2019, 2020; Yang and John 2012; Huang et al. 2008; Zavadskas et al. 2009; Haq and Kannan 2007; Lin and Lee 2007; Lin et al. 2008). These distinct forms of grey numbers not only capable at dealing with the presence of uncertainty in the heterogeneous preferences elicitation but also demonstrate great capability when they complement many decision making problems. Among other are supply chain management (Haq and Kannan 2007), forecasting (Lin and Lee 2007), software effort estimation (Huang et al. 2008), subcontractor selection (Lin et al. 2008) and contractor's selection (Zavadskas et al. 2009). For this reason, this chapter introduces for the first time a novel fuzzy risk assessment method based on Z-numbers with incorporation of grey numbers that is developed to ensure the presence of uncertainty in the heterogeneous preferences elicited by the risk analysts is holistically acknowledged. As part of the methodology, this study presents a novel fuzzy agreement relation approach that aims at handling the interactions between the common and the uncommon heterogeneous preferences elicited by the risk analysts. Then, results obtained from this step are aggregated using a novel fuzzy risk evaluation rating method to assess the correct level of risks, such that the assessments are consistent with the presence of uncertainty in the heterogeneous preferences elicited by the risk analysts. Later on, this study presents a real-world fuzzy risk assessment problem in electrical arc welding industry to validate the novelty and feasibility of the proposed methodology.

The rest of the chapter is structured as follows. Section 4.2 introduces the theoretical preliminaries related to this study. Section 4.3 presents the proposed novel fuzzy risk assessment method based on Z-numbers with incorporation of grey numbers. Section 4.4 covers the theoretical validation of the proposed method. Section 4.5 illustrates the numerical validation by means of application of the proposed novel fuzzy risk assessment method based on Z-numbers with incorporation of grey number on a real-world risk assessment problem in the electrical arc welding industry. Finally, discussion and conclusion are given in Sects. 4.6 and 4.7, respectively.

4.2 Literature Review

4.2.1 Z-Number

As overcoming the uncertainty in human decision making is crucial, the concept of Z-numbers (Zadeh 2011) is introduced by incorporating the element of reliability along with the decision restriction. This concept enhances the established concepts of type-1 fuzzy numbers and type-2 fuzzy numbers, where both consider uncertain decision with confidence level (Bakar and Gegov 2014) and inter-intra uncertainty (Bakar et al. 2019; Jana and Ghosh 2018; Wallsten and Budescu 1995; Yaakob et al. 2015; John and Coupland 2009), respectively. Based on (Zadeh 2011), the definition of Z-number is given as the following Definition 1.

Definition 1 (Zadeh 2011) A Z-number is an ordered pair of type-1 fuzzy numbers denoted as $Z = (A, B)$. The first component, A, is known as the restriction component where it is a real-valued uncertain on X whereas the second component, B, is the measure of reliability for A, presented as Fig. 4.1.

With respect to application of Z-numbers in fuzzy risk assessment, risks are represented as an ordered pair of risk restriction and the reliability of the restriction (Jiang et al. 2017; Wu et al. 2018; Abiyev et al. 2018). This can be seen when Z-numbers complement risk assessment problems in the literature such as risk assessment evaluations in failure mode of rotor blades of an aircraft turbine (Jiang et al. 2017), investigation on risk components in manufacturing and medical industries (Wu et al. 2018) and assessment of risk in food security (Abiyev et al. 2018).

$$\mu_A(x)=(a_1,a_2,a_3,a_4)=\begin{cases}\dfrac{x-a_1}{a_2-a_1} & if \quad a_1\le x\le a_2\\ 1 & if \quad a_2\le x\le a_3\\ \dfrac{a_3-x}{a_4-a_3} & if \quad a_3\le x\le a_4\\ 0 & otherwise\end{cases} \qquad \mu_B(x)=(b_1,b_2,b_3,b_4)=\begin{cases}\dfrac{x-b_1}{b_2-b_1} & if \quad b_1\le x\le b_2\\ 1 & if \quad b_2\le x\le b_3\\ \dfrac{b_3-x}{b_4-b_3} & if \quad b_3\le x\le b_4\\ 0 & otherwise\end{cases}$$

Fig. 4.1 A Z-number

4.2.2 Grey Number

The concept of grey numbers is introduced in the literature as to acknowledge the presence of non-homogeneous decision makers' preferences that are completely known, partially known, completely unknown and partially unknown (Bakar et al. 2020, 2019; Yang and John 2012; Huang et al. 2008). Definition of grey number and its further extensions are given as follows.

Definition 2 (Yang and John 2012) A grey number, G_A, is a number with clear upper and lower boundaries but has an unknown position within the boundaries. Mathematically, a grey number for the system is expressed as

$$G_A \in \left[g^-, g^+\right] = \left\{g^- \le t \le g^+\right\} \tag{4.1}$$

where t is information about $g^{\pm}$ while g^- and g^+ are the upper and lower limits of information t, respectively.

Definition 3 (Bakar et al. 2020; Yang and John 2012) For a set $A \subseteq U$, if its membership function value of each x with respect to A, $g_A^{\pm}(x)$, can be expressed with a grey number, $g_A^{\pm}(x) \in \bigcup_{i=1}^{n}\left[a_i^-, a_i^+\right] \in D[0,1]^{\pm}$, then A is a grey set, where $D[0,1]^{\pm}$ is the set of all grey numbers within the interval $[0, 1]$.

Definition 4 (*White Sets*) For a set $A \subseteq U$, if its membership function value of each x with respect to A, $g_{A_i}^{\pm}(x), i = 1, 2, \ldots, n$, can be expressed with a white number, then A is a white set.

Definition 5 (*Black Sets*) For a set $A \subseteq U$, if its membership function value of each x with respect to A, $g_{A_i}^{\pm}(x), i = 1, 2, \ldots, n$, can be expressed with a black number, then A is a black set.

Definition 6 (*Grey Sets*) For a set $A \subseteq U$, if its membership function value of each x with respect to A, $g_{A_i}^{\pm}(x), i = 1, 2, \ldots, n$, can be expressed with a grey number, then A is a grey set.

The following Table 4.1 presents comparison between white number, black number and grey number.

Established fuzzy risk assessment methods based on Z-numbers capable at dealing with the presence of uncertainty (Marhamati et al. 2018; Peng et al. 2019; Hendiani

Table 4.1 Comparison between white number, black number and grey number (Baker et al. 2019, 2020)

Number	Description	Value form	Preference elicitation form
0	Black number	Numerical	Completely unknown
[0, 1]	Grey Number	Interval	Partially known/unknown
1	White number	Numerical	Completely known

et al. 2020; Azadeh and Kokabi 2016) but the presence of uncertainty the risk faced is not well acknowledged. This is depicted when they consider only the presence of uncertainty when preferences elicited by the risk analysts are partially known (Jiang et al. 2017; Wu et al. 2018; Abiyev et al. 2018). Nonetheless, the presence of uncertainty can also happen when preferences elicited by the risk analysts are completely known, completely unknown and partially unknown (Bakar et al. 2020; Yang and John 2012; Huang et al. 2008). This points out that established fuzzy risk assessment methods based on Z-numbers do not have the holistic feature as they restrict the presence of uncertainty in the preferences elicited by the risk analysts to be homogeneous (partially known only), even if the presence of uncertainty is actually heterogeneous in nature (Bakar et al. 2020). Apart from that, the interactions between the common and uncommon heterogeneous preferences elicited by the risk analysts also indicate that the established fuzzy risk assessment methods based on Z-numbers are unable to holistically track the performance of risks in the presence of uncertainty. The above-mentioned inefficiencies of the established fuzzy risk assessment methods based on Z-numbers point out the motivations for this study. In order to resolve the mentioned limitations of established fuzzy risk assessment methods based on Z-numbers, this study utlises the concept of grey number (Yang and John 2012; Bakar et al. 2019) to acknowledge the presence of uncertainty when preference elicited by the risk analysts are heterogeneous. This concept is suitable for this study because it defines the completely known preferences elicitation in the form of white number, partially known and unknown preferences elicitation as the grey number and completely unknown preferences elicitation as the black number (Baker et al. 2019, 2020; Yang and John 2012; Huang et al. 2008; Zavadskas et al. 2009; Haq and Kannan 2007; Lin and Lee 2007; Lin et al. 2008).

4.3 Method

As mentioned in the introduction section, established fuzzy risk assessment methods based on Z-numbers do not holistically acknowledge the presence of uncertainty in the preferences elicited by the risk analysts. They consider the presence of uncertainty only when the preferences elicited by the risk analyst's is partially known (homogeneous), even if the presence of uncertainty can also happen when preferences elicited by the risk analysts are completely known, completely unknown and partially unknown (Baker et al. 2019, 2020; Yang and John 2012; Huang et al. 2008). Thus, for this reason, this study first incorporates the concept of grey numbers into Z-numbers, by means of describing both restriction and reliability components of the Z-numbers, as the grey numbers. This is to ensure that fuzzy risk assessment involving the presence of uncertainty in the heterogeneous preferences elicited by the risk analysts is holistically acknowledged. In other words, fuzzy risk assessment involving the presence of uncertainty when the preferences elicited by the risk analysts that are completely known, completely unknown, partially known and partially unknown are allowed to be existed. Since, the incorporation of grey numbers into Z-numbers,

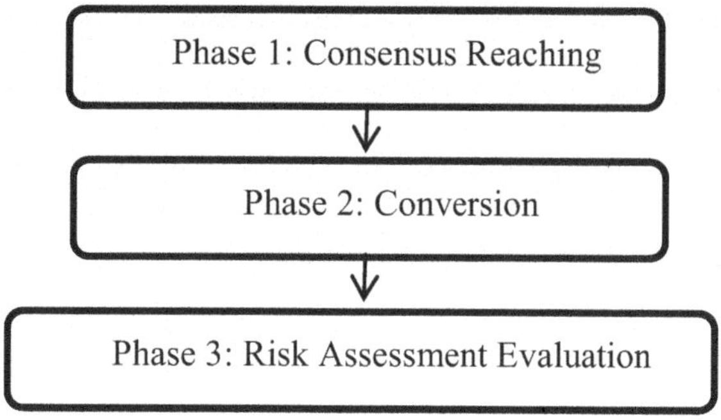

Fig. 4.2 Structure of the proposed method

Z-grey numbers, is developed for the first time here, this study defines the Z-grey number by the following definition.

Definition 3.1 (*Z-grey number*) Based on Definition 1, the Z-grey number is an ordered pair of grey number denoted as $Z_G = (G_A, G_B)$ with the first component, G_A, is known as the restriction component whereas the second component, G_B, is the measure of reliability for G_A.

It is worth noting that the Z-grey number is a Z-number with both restriction and reliability components are grey numbers. The incorporation of grey numbers into Z-numbers in this definition is consistent with Definition 1 (Bakar et al. 2020, 2019). With respect to fuzzy risk assessment, a novel fuzzy risk assessment method based on Z-grey numbers is proposed for the first time to resolve the inefficiencies faced by the established fuzzy risk assessment methods based on Z-numbers. This novel fuzzy risk assessment method consists of three phases, namely, consensus reaching phase, conversion phase and risk assessment evaluation phase. Details on each phase are given as follows (Fig. 4.2).

4.3.1 Phase 1: Consensus Reaching

In this phase, fuzzy risk assessment involving the presence of uncertainty in the heterogeneous preferences elicited by the risk analysts is defined in the form of Z-grey numbers. As both restriction of the preferences elicited by the risk analysts and the reliability of the restriction are grey numbers, the fuzzy risk assessment representations can exist in the form of white numbers (for completely known risk analysts' preferences elicitation), black numbers (completely unknown risk analysts' preferences elicitation) and grey numbers (partially known/unknown risk analysts' preferences elicitation). Since, the grey numbers forms are distinct from one to another (as in Table 4.1), a novel fuzzy agreement relation approach is developed for the first time to define fuzzy risk assessment representations involving the presence of uncertainty for each heterogeneous form of preferences elicited by the risk analysts as a single common form. The approach involves the transformation of Z-grey number into Z-number, where the transformation is given as follows.

Let $P'_{S_{i,k}}$ and $Q'_{S_{i,k}}$ be the probability of failure and the severity of loss, respectively, in the form of Z-grey numbers define as $P'_{S_{i,k}} = \left[H^G_{P'_{S_{i,k}}}, L^G_{P'_{S_{i,k}}}\right]$ and $Q'_{S_{i,k}} = \left[H^G_{Q'_{S_{i,k}}}, L^G_{Q'_{S_{i,k}}}\right]$, where $H^G_{P'_{S_{i,k}}}$ and $H^G_{Q'_{S_{i,k}}}$ are the restriction of the preferences elicited by the risk analysts for $P'_{S_{i,k}}$ and $Q'_{S_{i,k}}$ respectively, while $L^G_{P'_{S_{i,k}}}$ and $L^G_{Q'_{S_{i,k}}}$ are the reliability of the restriction for $P'_{S_{i,k}}$ and $Q'_{S_{i,k}}$, respectively.

1. If $P'_{S_{i,k}} \in [0, 1]$ and $Q'_{S_{i,k}} \in [0, 1]$ are Z-grey numbers that represent the preferences elicited by the risk analysts that are completely known, then $P'_{S_{i,k}}$ and $Q'_{S_{i,k}}$ are transformed into Z-numbers, $P^*_{S_{i,k}}$ and $Q^*_{S_{i,k}}$, respectively using the transformation function, T_σ, $\sigma = P'_{S_{i,k}}, Q'_{S_{i,k}}$, given as the following Eqs. (4.2) and (4.3).

$$T_{P'_{C_{i,k}}} : [0, 1] \to P^*_{S_{i,k}} = \left[H^G_{P^*_{S_{i,k}}}, L^G_{P^*_{S_{i,k}}}\right] \tag{4.2}$$

and

$$T_{Q'_{C_{i,k}}} : [0, 1] \to Q^*_{S_{i,k}} = \left[H^G_{Q^*_{S_{i,k}}}, L^G_{Q^*_{S_{i,k}}}\right] \tag{4.3}$$

2. If $P'_{S_{i,k}} \in [0, 1]$ and $Q'_{S_{i,k}} \in [0, 1]$ are Z-grey numbers that represent the preferences elicited by the risk analysts that are completely unknown, then $P'_{S_{i,k}}$ and $Q'_{S_{i,k}}$ are transformed into Z-numbers, $P^*_{S_{i,k}}$ and $Q^*_{S_{i,k}}$, respectively using the transformation function, T_ν, $\nu = P'_{S_{i,k}}, Q'_{S_{i,k}}$, given as the following Eqs. (4.4) and (4.5).

$$T_{P'_{C_{i,k}}} : [0, 1] \to P^*_{S_{i,k}} = \left[H^G_{P^*_{S_{i,k}}}, L^G_{P^*_{S_{i,k}}}\right] \tag{4.4}$$

and

$$T_{Q'_{C_{i,k}}} : [0, 1] \to Q^*_{S_{i,k}} = \left[H^G_{Q^*_{S_{i,k}}}, L^G_{Q^*_{S_{i,k}}}\right] \tag{4.5}$$

3. If $P'_{S_{i,k}} \in [0, 1]$ and $Q'_{S_{i,k}} \in [0, 1]$ are Z-grey numbers that represent the preferences elicited by the risk analysts that are partially known or partially unknown, then $P'_{S_{i,k}}$ and $Q'_{S_{i,k}}$ are transformed into Z-numbers, $P^*_{S_{i,k}}$ and $Q^*_{S_{i,k}}$, respectively using the transformation function, T_δ, $\delta = P'_{S_{i,k}}, Q'_{S_{i,k}}$, given as the following Eqs. (4.6) and (4.7).

$$T_{P'_{C_{i,k}}} : [0, 1] \to P^*_{S_{i,k}} = \left[H^G_{P^*_{S_{i,k}}}, L^G_{P^*_{S_{i,k}}}\right] \tag{4.6}$$

and

$$T_{Q'_{C_{i,k}}} : [0\ ,\ 1] \rightarrow Q^*_{S_{i,k}} = \left[H^G_{Q^*_{S_{i,k}}}\ ,\ L^G_{Q^*_{S_{i,k}}}\right] \tag{4.7}$$

4.3.2 Phase 2: Conversion

Note that from Phase 1, the current form for the fuzzy risk assessment representations involving the presence of uncertainty in the heterogeneous preferences elicited by the risk analysts is a single consensus form, which is the Z-numbers. The representation of the obtained single consensus form, however, is too complex in nature (Bakar and Gegov 2015; Kang et al. 2012; Zadeh 2011). Thus in this phase, this study converts the obtained single consensus form into a much simpler consensus form, which is the Z-fuzzy number. The conversion which involves incorporation of defuzzified value of the risk reliability into the risk restriction component, converts the obtained single consensus form (Z-numbers) $P^*_{S_{i,k}}$ and $Q^*_{S_{i,k}}$ into the reduced consensus form (Z-fuzzy numbers), $P^o_{S_{i,k}}$ and $Q^o_{S_{i,k}}$, respectively (Bakar and Gegov 2015; Kang et al. 2012). Details on the conversion are given by the following procedures (Bakar and Gegov 2015; Kang et al. 2012).

Step 1: Obtain the defuzzified value, T_n, of $L^G_{P^*_{S_{i,k}}}$ and $L^G_{Q^*_{S_{i,k}}}$ for both $P^*_{S_{i,k}}$ and $Q^*_{S_{i,k}}$, respectively, using the following Eq. (4.8).

$$T_n = \frac{1}{3}\left[b_{n1} + b_{n2} + b_{n3} + b_{n4} - \frac{b_{n3}\, b_{n4} - b_{n1}\, b_{n2}}{(b_{n3} + b_{n4}) - (b_{n1} + b_{n2})}\right] \tag{4.8}$$

where $n = L^G_{P^*_{S_{i,k}}}\ ,\ L^G_{Q^*_{S_{i,k}}}$.

Step 2: Incorporate T_n into $H^G_{P^*_{S_{i,k}}}$ and $H^G_{Q^*_{S_{i,k}}}$ for both $P^*_{S_{i,k}}$ and $Q^*_{S_{i,k}}$, respectively, using the following Eq. (4.9).

$$X_m = [T_n * a_{m1}, T_n * a_{m2}, T_n * a_{m3}, T_n * a_{m4};\ 1] = \left[\tilde{a}_{m1}, \tilde{a}_{m2}, \tilde{a}_{m3}, \tilde{a}_{m4};\ 1\right] \tag{4.9}$$

where $X = P^o_{S_{i,k}}, Q^o_{S_{i,k}}$ and $m = H^G_{P^*_{S_{i,k}}}\ ,\ H^G_{Q^*_{S_{i,k}}}$.

4.3.3 Phase 3: Risk Assessment Evaluation

In phase 2, fuzzy risk assessment representations involving the presence of uncertainty in the heterogeneous preferences elicited by the risk analysts in the form of

Z-grey number, has successfully converted into the reduced consensus forms (Z-fuzzy numbers). This reduced consensus forms are then aggregated using a novel fuzzy risk evaluation rating method to assess the correct level of risks, such that the assessments are consistent with the presence of uncertainty in the heterogeneous preferences elicited by the risk analysts. Steps provided in this phase are similar to established methods (Bakar and Gegov 2014, 2015; Baker et al. 2019, 2020), only that the proposed novel method uses Z-grey numbers. Details on the proposed novel fuzzy risk evaluation rating method are given as the following.

Step 1: Evaluate the interaction score, S_i, between $P^o_{S_{i,k}}$ and $Q^o_{S_{i,k}}$ for each risk under consideration as

$$S_i = \frac{\sum\limits_{i,k=1}^{n} \left(P^o_{S_{i,k}} \times Q^o_{S_{i,k}}\right)}{\sum\limits_{i,k=1}^{n} \left(Q^o_{S_{i,k}}\right)} \tag{4.10}$$

Step 2: Compute the centroid-x component value for S_i as

$$x_{S_i} = \frac{1}{3}\left[a_{1S_i} + a_{2S_i} + a_{3S_i} + a_{4S_i} - \frac{a_{3S_i}\,a_{4S_i} - a_{1S_i}\,a_{2S_i}}{\left(a_{3S_i} + a_{4S_i}\right) - \left(a_{1S_i} + a_{2S_i}\right)}\right] \tag{4.11}$$

and the centroid-y component value for S_i as

$$y_{S_i} = \frac{w_{S_i}}{3}\left[1 + \frac{a_{3S_i}\,a_{4S_i} - a_{1S_i}\,a_{2S_i}}{\left(a_{3S_i} + a_{4S_i}\right) - \left(a_{1S_i} + a_{2S_i}\right)}\right] \tag{4.12}$$

where $x_{S_i} \in [0, 1]$ and $y_{S_i} \in [0, 1]$.

Step 3: Obtain the deviation of centroid component value for S_i as

$$\psi_{S_i} = \left| a_{4S_i} - a_{1S_i} \right| \times y_{S_i} \tag{4.13}$$

Step 4: Evaluate the risk assessment evaluation rating value for all S_i under consideration as

$$A_{S_i} = x_{S_i} \times y_{S_i} \times \left(1 - \psi_{S_i}\right) \tag{4.14}$$

Risk assessment evaluation rating value descriptions:

If $A_{S_i} > A_{S_j}$, then $S_i(x) \succ S_j(x)$.
If $A_{S_i} = A_{S_j}$, then $S_i(x) \approx S_j(x)$.
If $A_{S_i} < A_{S_j}$, then $S_i(x) \prec S_j(x)$.

4.4 Theoretical Validation

In this section, the proposed novel fuzzy risk assessment method based on Z-grey numbers is theoretically validated based on ranking fuzzy quantity (Bakar and Gegov 2015; Baker et al. 2019, 2020). This validation serves as the generic analysis for risk assessment evaluations made by the proposed method in distinguishing which risk is riskier than other risks under consideration. Details on the validation are given as follows.

Let Z_A and Z_B be risk *A* and risk *B*, respectively, in the form of Z-grey numbers. Meanwhile, A_{Z_A} and A_{Z_B} be the risk assessment evaluation for risk *A* and risk *B*, respectively, using the proposed novel fuzzy risk assessment method based on Z-grey numbers.

Property 1 *If* $Z_A \succcurlyeq Z_B$ *and* $Z_B \succcurlyeq Z_A$, *then* $Z_A \approx Z_B$.

Proof $Z_A \succcurlyeq Z_B$ implies that $A_{Z_A} \geq A_{Z_B}$ and $Z_B \succcurlyeq Z_A$ implies that $A_{Z_B} \geq A_{Z_A}$, thus $A_{Z_A} = A_{Z_B}$ which is $Z_A \approx Z_B$.

Property 2 *If* $Z_A \succcurlyeq Z_B$ *and* $Z_B \succcurlyeq Z_C$, *then* $Z_A \succcurlyeq Z_C$.

Proof $Z_A \succcurlyeq Z_B$ implies that $A_{Z_A} \geq A_{Z_B}$ and $Z_B \succcurlyeq Z_C$ implies that $A_{Z_B} \geq A_{Z_C}$, thus $A_{Z_A} \geq A_{Z_C}$ which is $Z_A \succcurlyeq Z_C$.

Property 3 *If* $Z_A \cap Z_B = \varphi$ *and* Z_A *is on the right side of* Z_B, *then* $Z_A \succcurlyeq Z_B$.

Proof $Z_A \cap Z_B = \varphi$ and Z_A is on the right side of Z_B implies that $A_{Z_A} \geq A_{Z_B}$, thus $Z_A \succcurlyeq Z_B$.

Property 4 *The order of* Z_A and Z_B *are not affected by other Z-grey numbers under comparison.*

Proof The ordering of Z_A and Z_B are completely determined by A_{Z_A} and A_{Z_B} respectively, thus the ordering of Z_A and Z_B are not affected by other Z-grey numbers under comparison.

4.5 Case Study: Fuzzy Risk Assessment in Electrical Arc Welding

Consider the following real-world risk assessment problem experienced by a welding factory, which is the electrical arc welding. In the factory, risk assessment has become one of the most crucial aspects considering the presence of multiple types of hazards that may affect the safety of the workers during the operation of the electrical arc welding. Among the hazardous situations that involve in the electrical arc welding operations are exposure towards flammable substances, welding on wet floor, inhales

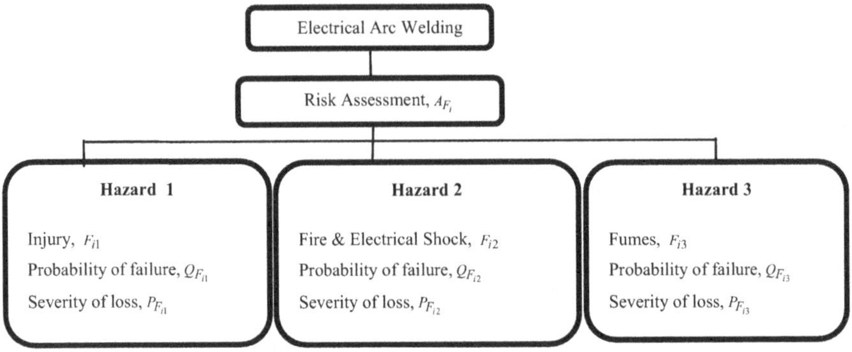

Fig. 4.3 Structure of risk assessment on the hazardous situations in the electrical arc welding operation

toxic welding fumes and least protection towards extreme bright flash. To ensure that the safety of the factory workers is well-supervised, the risks of all of the mentioned hazardous situations have to be assessed. The following are the hazardous situations that involve in the electrical arc welding operations and their descriptions.

1. Injury—Radiation that burn the workers' skin; extremely bright flash that damages the workers' eyes.
2. Fire and Electrical Shock—Exposure towards flammable substances (paper and thinner) when the welding process is carried out; exposure towards electrical shock when the floor is wet.
3. Fumes—Workers inhale toxic welding fumes created from the electrical arc process.

Based on these details, the structure of risk assessment for the electrical arc welding operations is illustrated as Fig. 4.3.

In the following, the proposed fuzzy risk assessment method based on Z-grey numbers that is developed in Sect. 4.3 is applied to assess the correct risk ordering for all hazardous situations under consideration, such that the ordering result is consistent with the actual risk evaluation on the level of hazards in the electrical arc welding operations. The actual risk information of by each risk analyst in the form Z-grey numbers is given in Table 4.2, while details on the proposed fuzzy risk assessment method based on Z-grey numbers are presented as follows.

Phase 1: Consensus Reaching

The transformation of actual risk information of by each risk analyst in the form Z-grey numbers into Z-number is given in Table 4.3.

Phase 2: Conversion

The transformation of actual risk information of by each risk analyst in the form of Z-numbers into Z-fuzzy number is given in Table 4.4.

Table 4.2 Actual risk information in the form of Z-grey numbers

Hazard	Risk analyst	Risk information (Z-grey numbers)	
		Severity of loss	Probability of failure
F_1 (injury)	F_{11}	$P_{C_{11}} = [0.58\,,\ (0.93, 0.98)]$	$Q_{C_{11}} = [0.92\,,\ (0.78\,;0.9)]$
	F_{12}	$P_{C_{12}} = [(0.10, 0.14)\,,\ (0.92, 0.97)]$	$Q_{C_{12}} = [(0.58\,, 0.63),\ (0.98, 1.0\,;0.7)]$
	F_{13}	$P_{C_{13}} = [0.78\,,\ 0.93]$	$Q_{C_{13}} = [0.8\,,\ (0.97\,;0.8)]$
F_2 (Fire and electrical shock)	F_{21}	$P_{C_{21}} = [1.0\,,\ (1.0, 1.0)]$	$Q_{C_{21}} = [0.93\,,\ (0.93\,;0.85)]$
	F_{22}	$P_{C_{22}} = [(0.78, 0.92)\,,\ 0.93]$	$Q_{C_{22}} = [(0.58, 0.63)\,,\ (0.98\,;0.95)]$
	F_{23}	$P_{C_{23}} = [(0.63, 0.80)\,,\ 0.80]$	$Q_{C_{23}} = [(0.72, 0.78)\,,\ (0.93, 1.0\,;0.9)]$
F_3 (fumes)	F_{31}	$P_{C_{31}} = [0.22\,,\ 0.97]$	$Q_{C_{31}} = [(0.80, 0.86)\,,\ (0.97\,;0.95)]$
	F_{32}	$P_{C_{32}} = [0.86\,,\ 0.86]$	$Q_{C_{32}} = [0.92\,,\ (0.92\,;0.8)]$
	F_{33}	$P_{C_{33}} = [(0.18, 0.23)\,,\ 0.72]$	$Q_{C_{33}} = [0.80\,,\ (0.93\,;1.0)]$

Phase 3: Risk Assessment Evaluation

Step 1: The interaction scores of the hazardous situations under consideration, in the form of Z-fuzzy numbers, are calculated and tabulated in Table 4.5. The score represents the aggregated risk interaction score evaluation between probability of failure and severity of loss factors for each respective hazardous situation under consideration.

Step 2–3: The centroid-x, centroid-y and spread components for all hazardous situations under consideration are computed and presented in Table 4.6.

Step 4: The risk assessment evaluation rating for all hazardous situations are calculated and given in Table 4.7.

Based on Table 4.7, the proposed novel fuzzy risk assessment method based on Z-grey numbers evaluates F_2 as the most hazardous situation, followed by F_1 and then F_3. In this case, fire and electrical shock hazard is assessed as the most hazardous situation in the electrical arc welding operations as compared to injury and fumes, because it has the highest risk assessment evaluation rating.

4.6 Discussion

In order to validate the novelty and feasibility of the proposed novel fuzzy risk assessment method based on Z-grey numbers, this study analyses the risk assessment evaluation rating results obtained with the actual risk evaluation on the level of hazards in the electrical arc welding operations. It is worth mentioning that, the

Table 4.3 Risk information in the form of Z-numbers

Hazard	Risk analyst	Risk information (Z-numbers)	
		Severity of loss	Probability of failure
F_1	F_{11}	$P_{C_{11}} = \begin{bmatrix} (0.93, 0.98, 1.0, 1.0), \\ (0.93, 0.98, 1.0, 1.0) \end{bmatrix}$	$Q_{C_{11}} = \begin{bmatrix} (0.17, 0.22, 0.36, 0.42), \\ (0.72, 0.78, 0.92, 0.97\,; 0.9) \end{bmatrix}$
	F_{12}	$P_{C_{12}} = \begin{bmatrix} (0.58, 0.63, 0.8, 0.86), \\ (0.72, 0, 78, 0.92, 0.97) \end{bmatrix}$	$Q_{C_{12}} = \begin{bmatrix} (0.32, 0.41\,, 0.58, 0.6), \\ (0.93, 0.98, 1.0, 1.0\,; 0.7) \end{bmatrix}$
	F_{13}	$P_{C_{13}} = \begin{bmatrix} (0.04, 0.1, 0.18, 0.23), \\ (0.93, 0.98, 1.0, 1.0) \end{bmatrix}$	$Q_{C_{13}} = \begin{bmatrix} (0.58, 0.63, 0.8, 0.86), \\ (0.72, 0.78, 0., 92, 0.97\,; 0.8) \end{bmatrix}$
F_2	F_{21}	$P_{C_{21}} = \begin{bmatrix} (0.04, 0.1, 0.18, 0.23), \\ (0.93, 0.98, 1.0, 1.0) \end{bmatrix}$	$Q_{C_{21}} = \begin{bmatrix} (0.93, 0.98, 1.0, 1.0), \\ (0.93, 0.98, 1.0, 1.0\,; 0.85) \end{bmatrix}$
	F_{22}	$P_{C_{22}} = \begin{bmatrix} (0.58, 0.63, 0.8, 0.86), \\ (0.93, 0.98, 1.0, 1.0) \end{bmatrix}$	$Q_{C_{22}} = \begin{bmatrix} (0.58, 0.63, 0.8, 0.86), \\ (0.93, 0.98, 1.0, 1.0\,; 0.95) \end{bmatrix}$
	F_{23}	$P_{C_{23}} = \begin{bmatrix} (0.72, 0.78, 0.92, 0.97), \\ (0.58, 0.63, 0.8, 0.86) \end{bmatrix}$	$Q_{C_{23}} = \begin{bmatrix} (0.58, 0.63, 0.80, 0.86), \\ (0.93, 0.98, 1.0, 1.0\,; 0.9) \end{bmatrix}$
F_3	F_{31}	$P_{C_{31}} = \begin{bmatrix} (0.17, 0.22, 0.36, 0.42), \\ (0.72, 0.78, 0.92, 0.97) \end{bmatrix}$	$Q_{C_{31}} = \begin{bmatrix} (0.17, 0.22, 0.36, 0.42), \\ (0.72, 0.78, 0.92, 0.97\,; 0.95) \end{bmatrix}$
	F_{32}	$P_{C_{32}} = \begin{bmatrix} (0.58, 0.63, 0.8, 0.86), \\ (0.58, 0.63, 0.8, 0.86) \end{bmatrix}$	$Q_{C_{32}} = \begin{bmatrix} (0.72, 0.78, 0.92, 0.97), \\ (0.72, 0.78, 0.92, 0.97\,; 0.8) \end{bmatrix}$
	F_{33}	$P_{C_{33}} = \begin{bmatrix} (0.04, 0.1, 0.18, 0.23), \\ (0.72, 0.78, 0.92, 0.97) \end{bmatrix}$	$Q_{C_{33}} = \begin{bmatrix} (0.58, 0.63, 0.80, 0.86), \\ (0.58, 0.63, 0.80, 0.86\,; 1.0) \end{bmatrix}$

Table 4.4 Risk information in the form of Z-fuzzy numbers

Hazard	Risk analyst	Risk information (Z-fuzzy numbers)	
		Severity of loss	Probability of failure
F_1	F_{11}	$P^*_{F_{11}} = (0.96, 0.98, 0.04, 0.00)$	$Q^*_{F_{11}} = (0.21, 0.35, 0.05, 0.06; 0.9)$
	F_{12}	$P^*_{F_{12}} = (0.53, 0.68, 0.05, 0.05)$	$Q^*_{F_{12}} = (0.35, 0.49, 0.08, 0.02; 0.7)$
	F_{13}	$P^*_{F_{13}} = (0.10, 0.18, 0.06, 0.05)$	$Q^*_{F_{13}} = (0.61, 0.78, 0.05, 0.06; 0.8)$
F_2	F_{21}	$P^*_{F_{21}} = (0.10, 0.18, 0.06, 0.05)$	$Q^*_{F_{21}} = (0.96, 0.98, 0.05, 0.00; 0.85)$
	F_{22}	$P^*_{F_{22}} = (0.61, 0.78, 0.05, 0.06)$	$Q^*_{F_{22}} = (0.61, 0.78, 0.05, 0.06; 0.95)$
	F_{23}	$P^*_{F_{23}} = (0.56, 0.66, 0.04, 0.04)$	$Q^*_{F_{23}} = (0.45, 0.57, 0.04, 0.04; 0.9)$
F_3	F_{31}	$P^*_{F_{31}} = (0.08, 0.15, 0.05, 0.04)$	$Q^*_{F_{31}} = (0.19, 0.31, 0.04, 0.05; 0.95)$
	F_{32}	$P^*_{F_{32}} = (0.45, 0.57, 0.04, 0.04)$	$Q^*_{F_{32}} = (0.56, 0.66, 0.04, 0.04; 0.8)$
	F_{33}	$P^*_{F_{33}} = (0.09, 0.15, 0.05, 0.04)$	$Q^*_{F_{33}} = (0.53, 0.68, 0.04, 0.05; 1.0)$

Table 4.5 Risk interaction score evaluation for each hazardous situation under consideration

Hazard	Risk interaction score evaluation
F_1	$F_1 = (0.54, 1.03, 0.13, 0.30; 0.7)$
F_2	$F_2 = (0.67, 0.94, 0.10, 0.14; 0.85)$
F_3	$F_3 = (0.39, 0.95, 0.13, 0.45; 0.8)$

Table 4.6 The centroid-x, centroid-y and spread components for all hazardous situations under consideration

Hazard	Component		
	Centroid-x	Centroid-y	Deviation of centroid
F_1	$x_{F_1} = 0.8301$	$y_{F_1} = 0.3150$	$\psi_{F_1} = 0.2893$
F_2	$x_{F_2} = 0.8115$	$y_{F_2} = 0.3812$	$\psi_{F_2} = 0.1937$
F_3	$x_{F_3} = 0.7579$	$y_{F_3} = 0.3553$	$\psi_{F_3} = 0.4049$

Table 4.7 The risk assessment evaluation rating for all hazardous situations under consideration

Hazard	Risk assessment evaluation rating
F_1	$A_{F_1} = 0.1858$
F_2	$A_{F_2} = 0.2495$
F_3	$A_{F_3} = 0.1603$

Table 4.8 Actual factory risk evaluation for all hazardous situations under consideration

Hazard	Injury, F_1	Fire and electrical shock, F_2	Fumes, F_3
Probability of failure	Moderate	Likely	Unlikely
Severity of loss	Moderate	Major	Minor
Level	Moderate	High	Low

Table 4.9 Risk assessment evaluation rating by risk assessment methods under consideration

Risk assessment Method	F_1	F_2	F_3	Risk assessment
Fuzzy numbers (Bakar and Gegov 2014)	0.4119	0.3997	0.5036	$F_3 \succ F_1 \succ F_2$
Z-numbers (Bakar and Gegov 2015)	0.1048	0.1839	0.1522	$F_2 \succ F_3 \succ F_1$
Grey numbers (Bakar et al. 2020)	0.8065	0.7565	0.8234	$F_3 \succ F_1 \succ F_2$
The proposed method	0.1858	0.2495	0.1603	$F_2 \succ F_1 \succ F_3$

actual risk evaluations are obtained from the factory risk assessment as shown in the following Table 4.8.

From the actual factory risk evaluations on all of the hazardous situations in Table 4.8, F_2 is considered as the most hazardous situation as it is the most likely to occur as compared to F_1 and F_3. Furthermore, the level of severity for F_2 is the highest from all of the hazardous situations under consideration. The company grades F_2 as high level of hazard but the most hazardous situation among those under consideration in this case. For F_1, the chance for the hazard to occur is moderate, meanwhile F_3 is unlikely to occur. With respect to levels of severity for F_1 and F_3, they are moderate and low, respectively. Thus, the company grades the level of hazard for all of the hazardous situations under consideration as $F_2 \succ F_1 \succ F_3$.

Methods Performance

Based on Table 4.9, only the proposed novel fuzzy risk assessment method based on Z-grey numbers obtains the correct level of risk for all hazardous situations under consideration such that the assessment result is consistent with the actual risk evaluation on the level of hazards in the electrical arc welding operations, i.e. $F_2 \succ F_1 \succ F_3$. This is due to the fact that the proposed method evaluates hazardous situation with the highest risk assessment evaluation rating as the most hazard situation. For comparative analysis purposes, this study compares the proposed novel fuzzy risk assessment method based on Z-grey numbers with established methods that utilise centroid point and deviation of centroid point but are applied in different forms.

Established methods (Bakar and Gegov 2014; Bakar et al. 2020) evaluate risk for each hazardous situation under consideration based on fuzzy numbers and gey numbers, respectively, where same risk assessment result is obtained, which is $F_3 \succ F_1 \succ F_2$. These methods assess F_3 as the most hazardous situation than F_1 and F_2. Their results are misleading as F_2 is graded as high risk by the factory, followed by F_1 and then F_3. In this respect, risk assessment evaluations by Bakar and Gegov

(2014) and Bakar et al. (2020) are considered to be incorrect such that the risk assessment results are inconsistent with the actual risk evaluation on the level of hazards in the electrical arc welding operations.

Established method (Bakar and Gegov 2015) calculates risk for each hazardous situation under consideration using Z-numbers as $F_2 \succ F_3 \succ F_1$. The method successfully grades F_2 as the most risky situation as compared to other hazardous situations under consideration. Nonetheless, the risk assessment evaluations obtained are incorrect for F_1 and F_3 as the latter is considered to be less risky than the former, even if F_1 is graded with higher risk level by the factory than F_3. Although, F_2 is correctly assessed, the overall risk assessment evaluation by Bakar and Gegov (2015) is still considered to be incorrect such that the risk assessment result is inconsistent with the actual risk evaluation on the level of hazards in the electrical arc welding operations. Thus, based on these risk assessment evaluations, the proposed novel fuzzy risk assessment method outperforms established fuzzy risk analysis methods under consideration.

4.7 Conclusion

In this chapter, a novel fuzzy risk assessment method based on Z-grey numbers has successfully developed. The main motivation of this study is to acknowledge the presence of uncertainty in the heterogeneous preferences elicited by the risk analysts, which is neglected by established fuzzy risk assessment methods based on Z-numbers. At the same time, the interactions between the common and uncommon heterogeneous preferences elicited by the risk analysts are also considered. In the proposed method, the establishment of the novel fuzzy agreement relation approach has complemented the presence of interactions between the common and the uncommon heterogeneous preferences elicited by the risk analysts. Other than that, the development of the novel fuzzy risk evaluation rating method has successfully assessed the correct level of risks, such that the assessment results are consistent with the presence of uncertainty in the heterogeneous preferences elicited by the risk analysts. Not only that, an application on real-world risk assessment problem in the electrical arc welding industry has also been demonstrated to signify the novelty, validity and feasibility of the proposed work. In the performance analysis conducted, the proposed novel fuzzy risk assessment method based on Z-grey number outperforms established fuzzy risk assessment methods based on Z-number under consideration. The main advantage of the proposed method is that it not only has the capability to acknowledge the presence of uncertainty in the heterogeneous preferences elicited by the risk analysts, but is also giving correct risk assessment evaluation such that the evaluation results are consistent with the actual preferences elicited by the risk analysts.

For future research, investigations on acknowledging the presence of uncertainty from the perspective of the preferences elicited by the risk analysts that are hesitant in nature, are to be carried out. This plan will complement many established and novel research with respect to acknowledging the presence of uncertainty in a more flexible and accurate way.

Acknowledgements This research work is funded by University of Malaya Research Grants GPF003H-2018 and the Malaysia Ministry of Education Research Grant FRGS/1/2018/STG06/UM/02/14.

References

Abiyev RH, Uyar K, Ilhan U, Imanov E, Abiyeva E (2018) Estimation of food security risk level using Z-number-based fuzzy system. J Food Qual 9

Allahviranloo T, Ezadi S (2019) Z-advanced numbers processes. Inf Sci 480:130–143

Azadeh A, Kokabi R (2016) Z-number DEA: a new possibilistic DEA in the context of Z-numbers. Adv Eng Inform 30(3):604–617

Bakar ASA, Gegov A (2014) Ranking of fuzzy numbers based on centroid point and spread. J Intell Fuzzy Syst 27(3):1179–1186

Bakar ASA, Gegov A (2015) Multi-layer decision making methodology for ranking Z-numbers. Int J Comput Intell Syst 8(2):395–406

Bakar ASA, Khalif KMNK, Yaakob AM, Gegov A, Amin AZM (2019) Grey number based methodology for non-homogeneous preference elicitation in fuzzy risk analysis management. Applying fuzzy logic for the digital economy and society Switzerland, pp 135–148. Springer

Bakar ASA, Khalif KMNK, Shariff AA, Gegov A, Salleh FM (2020) Fuzzy risk analysis under influence of non-homogeneous preferences elicitation in fiber industry. Appl Intell 50:157–168

Chukwuma EC, Rashid A, Okafor G, Nwoke AO (2020) Fuzzy based risk assessment of abattoir operations and treatment facilities: a case study of Onitsha North/South Lga of Anambra State of Nigeria. Food Bioprod Process 119:88–97

Haq N, Kannan G (2007) A hybrid normalised multi criteria decision making for the vendor selection in a supply chain model. Int J Manage Dec Making 8(5/6):601–622

Hendiani S, Bagherpour M, Mahmoudi A, Liao H (2020) Z-number based earned value management (ZEVM): a novel pragmatic contribution towards a possibilistic cost-duration assessment. Comput Ind Eng 143:106430

Huang SJ, Chiu NH, Chen LW (2008) Integration of the grey relational analysis with genetic algorithm for software effort estimation. Eur J Oper Res 188(3):898–909

Jana DK, Ghosh R (2018) Novel interval type-2 fuzzy logic controller for improving risk assessment model of cyber security. J Inf Secur Appl 40:173–182

Jiang W, Xie B, Wei B, Tang Y (2017) Failure mode and effects analysis based on Z-numbers. Intell Autom Soft Comput 24(1):165–172

John R, Coupland S (2009) Type-2 fuzzy logic and the modelling of uncertainty in applications. In: Bargiela A, Pedrycz W (eds) Human-centric information processing through granular modelling. Studies in computational intelligence, vol 182. Springer, Berlin

Kang B, Wei D, Li Y, Deng Y (2012) A method of converting Z-numbers to classical fuzzy numbers. J Inf Comput Sci 9(3):703–709

Lin YH, Lee PC (2007) Novel high-precision grey forecasting model. Autom Constr 16(6):771–777

Lin YH, Lee PC, Chang TP, Ting HI (2008) Multi-attribute group decision making model under the condition of uncertain information. Autom Constr 17(6):792–797

Marhamati N, Buxton EK, Rahimi S (2018) Integration of Z-numbers and Bayesian decision theory: a hybrid approach to decision making under uncertainty and imprecision. Appl Soft Comput 72:273–290

Natha Reddy VR, Gokulachandran J (2020) A comprehensive study in finding the risk assessment using a fuzzy analytical hierarchal model in textile industry: a case study. Mater Today Proc 24(2):895–903

Peng H, Wang X, Wang T, Wang J (2019) Multi-criteria game model based on the pairwise comparisons of strategies with Z-numbers. Appl Soft Comput 74:451–465

Wallsten TS, Budescu DV (1995) A review of human linguistic probability processing: general principles and empirical evidence. Knowl Eng Rev 10(1):43–62

Wu D, Liu X, Xue F, Zheng H, Shou Y, Jiang W (2018) A new medical diagnosis method based on Z-numbers. Appl Intell 48(4):854–867

Yaakob AM, Serguieva A, Gegov A (2015) FN-TOPSIS: fuzzy network for ranking traded equities. IEEE Trans Fuzzy Syst 25(2):315–332

Yang Y, John R (2012) Grey sets and greyness. Inf Sci 185(1):249–264

Zadeh LA (2011) A note on Z-numbers. Inf Sci 181:2923–2932

Zavadskas EK, Kaklauskas A, Turskis Z, Tamošaitiene J (2009) Multi-attribute decision making model by applying grey numbers. Informatica 20(2):305–320

Zhao B, Chen H, Gao D, Xu L (2020) Risk assessment of refinery unit maintenance based on fuzzy second generation curvelet neural network. Alex Eng J 59(3):1823–1831

Chapter 5
Assessment of COVID-19 Transmission Risk Through Fuzzy Inference System; an Application for Mining Activities

Ulas Cinar, Omer Faruk Ugurlu, and Selcuk Cebi

Abstract The current novel coronavirus has a huge impact on both human lives and the global economy. While on the one hand, vaccine research is being carried out, on the other hand, efforts to save the economy are being made. Mining activities are crucial for the global economy because they are one of the most important links in the supply chain. Therefore, various evaluations are required to continue mining activities without being affected by the pandemic. The aim of this study is to develop a method to effectively determine the current novel coronavirus transmission risk in mining activities for three main mining environments; open-pit, underground metal, and underground coal mines. The linguistic expressions made by the experts based on the literature were calculated by the Mamdani fuzzy inference system to establish the rule base. The rule base in the proposed method was used to analyze the risk of the current novel coronavirus transmission. It has been determined that the risk of transmission in mines dominated by mechanized production is within acceptable levels. In addition, the condition of the working environment is one of the factors affecting the transmission risk. The proposed method can be used to assess the risk of the current novel coronavirus transmission in mines. The results of the study are in line with the results made in different fields.

Keywords COVID-19 · Mamdani fuzzy inference system · Mining engineering · Occupational health and safety · Risk assessment

---

U. Cinar
Occupational Health and Safety Education Application and Research Center, Canakkale Onsekiz Mart University, Canakkale, Turkey
e-mail: ulas.cinar@comu.edu.tr

O. F. Ugurlu (✉)
Mining Engineering Department, Istanbul University-Cerrahpasa, Istanbul, Turkey
e-mail: ofugurlu@istanbul.edu.tr

S. Cebi
Industrial Engineering Department, Yildiz Technical University, Istanbul, Turkey
e-mail: scebi@yildiz.edu.tr

J. Ren (ed.), *Multi-Criteria Decision Analysis for Risk Assessment and Management*, Industrial Ecology and Environmental Management 1,
https://doi.org/10.1007/978-3-030-78152-1_5

5.1 Introduction

The current novel coronavirus (COVID-19) is a global pandemic that has caused infections and deaths all over the globe. People with weakened immune systems and over 40 are more vulnerable. The risk of serious illness increases with age and chronic diseases such as diabetes, heart, and lung diseases (WHO 2020a). The places where the virus has transmitted most are the workplaces. Therefore, personal hygiene and social distancing are the two key parameters to avoid COVID-19 transmission, particularly in the workplace (WHO 2020b).

This unpredicted and unprecedented outbreak has not only affected human lives, but has also wrecked the global economy (Ahamed and Samad 2004). The economy of many developed and developing countries directly depends on the activities in the mining sector. Therefore, mining activities must inevitably continue to keep the supply chain intact in the industry. However, the outbreak has a profound impact on the mining activities which are essential services. According to Fernandes (2020a), the mining sector has fallen by more than 30%. The demand for metals and minerals has decreased immensely. The reduction has caused extensive falls in the mineral prices and the production rate in the short term. These falls have been most dramatic for aluminum and copper (Laing 2019). The medium and long-term effects are highly uncertain (Baker et al. 2020); therefore, the risk assessment of virus transmission is vital to ensure that the mining sector can continue the operations.

The risk of transmission of the COVID-19 virus and its effects have only just begun to be understood, and the virus is still unknown. There have been a lot of studies conducted to explore the transmission characteristic of the virus (Hassen et al. 2020). COVID-19 often spreads by the droplets of infected fluids of someone who has coughed or even exhaled (Chen 2020). Meteorological conditions such as temperature, humidity, and ventilation speed have a crucial impact on the effect of the virus (Rosario et al. 2020). Touching contaminated surfaces and objects is one of the main reasons for the transmission of the virus (WHO 2019). Another reason is standing within a meter with an infected person (WHO 2020a). Mines are one of the environments with a high risk of COVID-19 transmission because mining activities often require large numbers of workers working, eating, sleeping, and bathing together in confined spaces. Social distancing is difficult and nearly impossible to practice in those conditions, contributing to increased risks of transmission. There is nothing more important than the safety and health of the workforce. Therefore, companies must adhere to strict preventive measures. While different companies have different measures and guidelines in place for businesses to operate through the pandemic such as reducing the production and workforce, social distancing measures, workplace hygiene policy, and temperature checks at the operations must be implemented (WHO 2020a).

Risk assessment is a crucial element for proactive safety management. Although risk assessment methods based on scoring are generally used to evaluate occupational health and safety risks, special methods have also been developed for the evaluation of particular areas such as chemicals that are stored together and machines with specific working principles (Cinar et al. 2020). It is evident that the assessment of problems which have negative results such as pandemic will be insufficient in reflecting the real risk since it is mostly expressed by using common linguistic methods. The main purpose of this study is to establish a rule base by compiling what is known about the transmission risk of the COVID-19 virus in the current literature and to develop a method depending on the rule base in order to effectively express COVID-19 transmission risk. Thus, the operations in the mining sector which have a high transmission risk will be identified and it will enable to develop more effective measures for these operations. A method was applied by a fuzzy inference system and applications were carried out in different production methods.

5.2 Literature Review

COVID-19 is a new phenomenon around the globe. There is a lot of research that has been carried out and most of them have been going on. It is expected to have accurate results in the near future. In this section, some researches related to the risk analysis and the fuzzy inference system are examined to show the eligibility of the method in order to measure the risk of COVID-19 transmission.

Rezaee et al. (2020) presented a hybrid approach based on the Linguistic FMEA, Fuzzy Inference System (FIS), and Fuzzy Data Envelopment Analysis (DEA) model to calculate a novel score for covering shortcomings and the prioritization of health, safety, and environmental risk factors in the chemical industry. The task of the fuzzy inference mechanism in this model is to remove the ambiguity in linguistic expressions and to transform complex data into meaningful outputs. Jamshidi et al. (2013) developed an application to assess pipeline risk using the Mamdani Fuzzy Inference System in engineering problems. The researchers aimed to integrate Relative Risk Score (RRS) methodology depending on the Mamdani algorithm with experts' knowledge. When compared with the evaluations made with classical methods, it has been observed that the proposed method gives more accurate and precise results.

Kim et al. (2016) conducted a study to provide valuable information regarding worker safety represented by a numerical accident analysis in dynamic environments such as construction sites. Firstly, computer vision was used to monitor a construction site and extract spatial information for each entity (workers and equipment). Then, a fuzzy inference system was used to assess the proper safety levels of each entity using spatial information. It was aimed to represent a safety level that shows the potential hazard or the integrating danger in the working environment.

A hybrid method including Fuzzy Inference System, Fuzzy AHP, and Fine Kinney methods was proposed by Ilbahar et al. (2018). Occupational health and safety risks were evaluated using the hybrid method. An application has been implemented in the construction industry using the Fuzzy Inference System to transform linguistic expressions into analytical data. It was aimed to provide a more accurate risk assessment in dynamic environments such as construction sites. The hybrid method and other methods were compared and the results showed that the hybrid method produced reliable and informative outcomes to represent better vagueness of the decision-making process. Similarly, Debnath et al. (2016) formulated a model to consider the risk factors and controlling factors for accidental injuries in construction sites. The Takagi–Sugeno Fuzzy Inference System was applied to the occupational health and safety risk assessment study recommended for the construction industry. In the model formulation process, the risk factors and controlling factors for accidental injuries were considered as input parameters. The applicability of the model was tested in the selected construction sites to validate the approach. Another study was conducted about the risk assessment of a construction project by using fuzzy systems (Ebrat and Ghodsi 2014). The authors designed to evaluate the risk of construction projects using the neuro-fuzzy inference system. The results of the study show that the model gives satisfactory information to practitioners.

Fuzzy Inference System was also used in the assessment of system safety and reliability. Ratnayake (2014) developed an optimization approach to simplify the failure risk of rotating equipment by having a more reliable estimation. In this study, the Fuzzy Inference System was used to minimize the suboptimal prioritizations of functions in the functional failure risk analysis using an illustrative tailor-made risk matrix. Moreover, Guimarães and Lapa (2007) presented a case study to analyze risk factors in nuclear platforms using the Fuzzy Inference System. Wang and Elhag (2008) developed a system to determine the maintenance priority of more than 500 bridge structures using neuro-fuzzy systems. The proposed method was compared with the existing bridge risk assessment methods to show the efficiency of the method. In addition, the fuzzy system based on the Mamdani inference can be used to perform an environmental risk assessment. Camastra et al. (2015) proposed a fuzzy decision approach to assess the environmental risk of the genetically modified plants.

The Fuzzy system and the Mamdani inference system have a wide range of uses as seen in the literature. Its use especially in construction sites has led to the idea that it can be used in mine production sites. Although mine production sites have similarities with construction sites, they actually have a very different environment; they differ even within themselves. The main difference of the proposed approach from other studies is that there is no observed study about mine production sites in the literature. In other words, there is no other study conducted to evaluate the risks in mining sites.

5.3 Methods

In the proposed method, the parameters affecting COVID-19 transmission risk in mining activities are determined as the number of employees, co-working time, co-working distance, and working environment for the production techniques. The literature studies about the COVID-19 were taken into consideration in determining the parameters and establishing the rule base for the mining activities (Liu et al. 2020). Each mining activity is weighted using the parameters by the Mamdani fuzzy inference system. The model characterizes a rule-based system, and the general structure of the system used in the model is given in Eq. (5.1) (Mamdani and Assilian 1999; Mamdani 1977).

$$\begin{aligned} &if\ x_1 = Z_{i1}\ and\ x_2 = Z_{i2}\ and\ x_3 = Z_{i3}\ and\ \ldots x_n = Z_{in} \\ &then\ y = P_i.\ i = 1, 2, 3, \ldots, k \end{aligned} \tag{5.1}$$

where x_n ($n = 1, 2, 3, \ldots m$) represents the input dataset, Z_i and P_i are linguistic expressions of membership function, y is the output value, and k is the number of rules in the rule base. If multiple discrete rules existing in the system are activated simultaneously, the result is usually obtained by using the max–min operator which is given in Eq. (5.2) (Mamdani and Assilian 1999; Mamdani 1977).

$$\mu_{Pk}(y) = maks[min[\mu_{Z1k}(x_1), \mu_{Z2k}(x_2)]],\ \ k = 1, 2, 3, \ldots, n \tag{5.2}$$

The μ_{pk}, μ_{Z1k}, and μ_{Z2k} given in the equation are the membership degrees of the y, x_1, and x_2, respectively. If there are more than one evaluator, the output value which is obtained as a fuzzy value from the model should be clarified. The centroid of area (also called center of gravity) method is used for the clarifying process which is given in Eqs. (5.3) and (5.4) (Mamdani and Assilian 1999; Mamdani 1977).

$$Z^*_{COZ} = \frac{\int_Z^x \mu_X(x)xdx}{\int_Z^x \mu_Z(x)dx} \tag{5.3}$$

$$Z^*_{COZ} = \frac{\sum_i^q \mu_Z(x_i)x_i}{\sum_i^q \mu_A(x_i)}\quad i = 1, 2, 3, \ldots, q \tag{5.4}$$

where Z^*_{COZ} is the exact value obtained from the system. More information about the Mamdani fuzzy inference system can be found in Ilbahar et al. (2018), Cinar and Cebi (2019), and Karasan et al. (2018).

5.4 Case Study

In this study, production methods in mines have been analyzed under three main mining environments (open-pit, underground metal, and underground coal mines) considering particularly the number of employees and the working condition. Open-pit mining is a mining technique of extracting valuable minerals by removing the soil from the earth (Ugurlu and Kumral 2019). It is an open-air activity in which many workers are needed. However, the co-working distance is generally large. Underground mining is another mining technique that is used to excavate hard minerals like metals and soft minerals such as coals and oil sands. Horizontal and vertical tunnels are created to access valuable minerals. The underground openings are large for metal mines. Most of the underground metal mines are mechanized; thus, the number of workers is limited and the underground openings are large. On the other hand, because of the geological structure, the underground coal mines in Turkey are not suitable for mechanized production. Hence, the underground openings of coal mines are relatively small and a large number of workers are needed.

The abbreviation and fuzzy numbers corresponding to the linguistic expressions used for the parameters in the proposed method and the linguistic evaluation scales for membership degrees are given in Tables 5.1, 5.2, 5.3, 5.4, 5.5 and Figs. 5.1, 5.2, 5.3, 5.4, 5.5.

The rules regarding the inference mechanism used in our study are given in Table 5.6 (Cinar et al. 2020).

Table 5.1 Linguistic expressions and trapezoidal fuzzy numbers (TrFNs) for the number of employees

Number of workers	Linguistic scale	Fuzzy				TrFns
1	Very low (VL)	1	1	2	3	(1, 1, 2, 3)
2–5	Low (L)	2	3	5	6	(2, 3, 5, 6)
6–10	Moderate (M)	5	6	9	10	(5, 6, 9, 10)
11–15	High (H)	9	10	13	15	(9, 10, 13, 15)
>15	Very high (VH)	13	15	17		(13, 15, 17)

Table 5.2 Linguistic expressions and fuzzy numbers for co-working time

Working time in cooperation (h)	Linguistic scale	Fuzzy				TrFns
<1/2	VL	0	0	0.25	0.5	(0, 0, 0.25, 0.5)
1/2–1	L	0.25	0.5	0.75	1	(0.25, 0.5, 0.75, 1)
1–2	M	0.75	1	1.5	2	(0.75, 1, 1.5, 2)
2–3	H	1.5	2	2.5	3	(1.5, 2, 2.5, 3)
>3	VH	2.5	3	5		(2.5, 3, 5)

Table 5.3 Linguistic expressions and fuzzy numbers for co-working distance

Types of working distance in cooperation (m)	Linguistic scale	Fuzzy				TrFns
<1	VL	0.5	0.5	0.75	1	(0.5, 0.5, 0.75, 1)
1–1.5	L	0.75	1	1.25	1.5	(0.75, 1, 1.25, 1.5)
1.5–2	M	1.25	1.5	1.75	2	(1.25, 1.5, 1.75, 2)
2–3	H	1.75	2	2.5	3	(1.75, 2, 2.5, 3)
>3	VH	2.5	3	5		(2.5, 3, 5)

Table 5.4 Linguistic expressions and fuzzy numbers for the working environment

Types of working environment	Linguistic scale	Fuzzy				TrFns
Open air (open-pit)	VL	0	0	0	1	(0, 0, 0, 1)
Underground, mechanical ventilation and large section	L	0	1	3	4	(0, 1, 3, 4)
Underground, mechanical ventilation and narrow section	M	3	4	6	7	(3, 4, 6, 7)
Underground, natural ventilation and large section	H	6	7	9	10	(6, 7, 9, 10)
Underground, natural ventilation and narrow section	VH	9	10	10	10	(9, 10, 10, 10)

Table 5.5 Linguistic expressions and fuzzy numbers for risk level

Linguistic terms	TrFns
Negligible (N)	(0, 0, 1, 3)
Minor (Mi)	(1, 3, 4, 6)
Major (Ma)	(4, 6, 7, 9)
Catastrophic (C)	(7, 9, 10, 10)

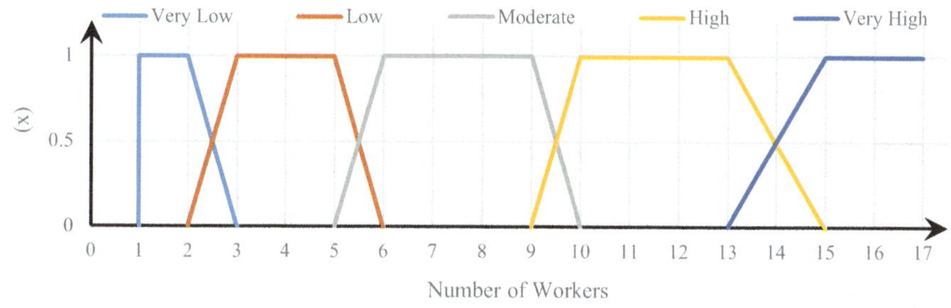

Fig. 5.1 Linguistic evaluation scales and membership degrees for the number of employees

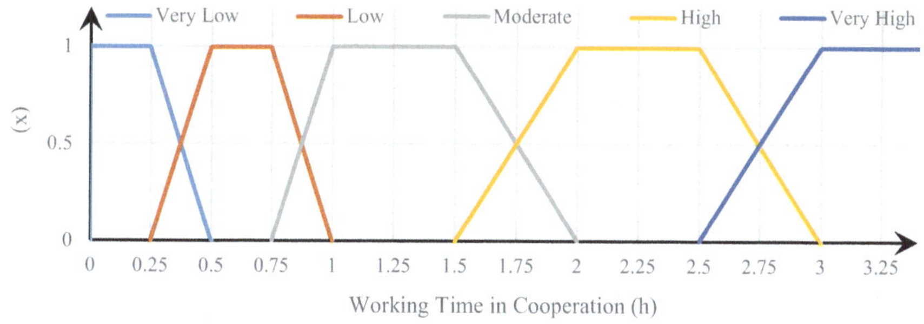

Fig. 5.2 Linguistic evaluation scales and membership degrees for co-work time

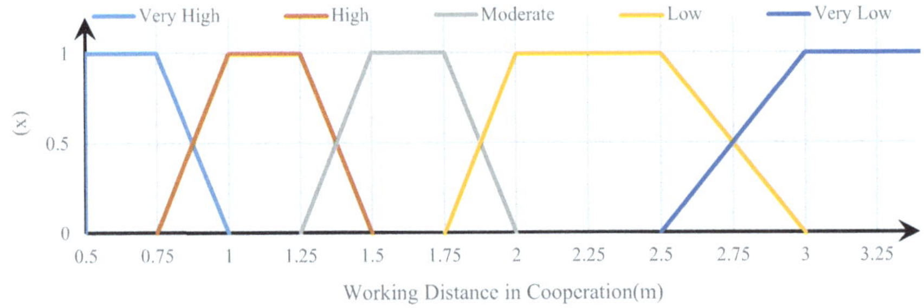

Fig. 5.3 Linguistic assessment scales and membership degrees for co-working distance

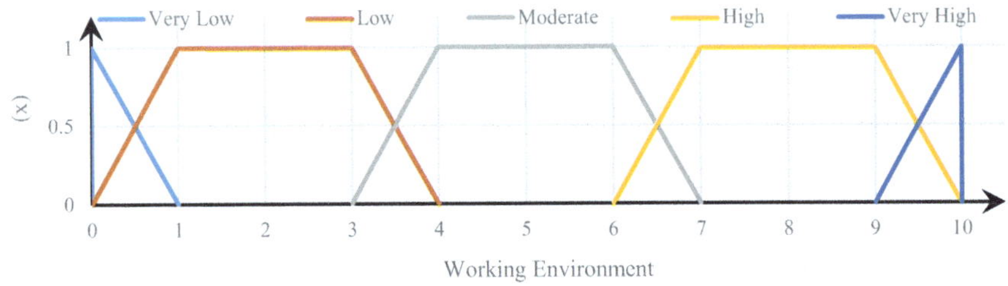

Fig. 5.4 Linguistic assessment scales and membership degrees for the work environment

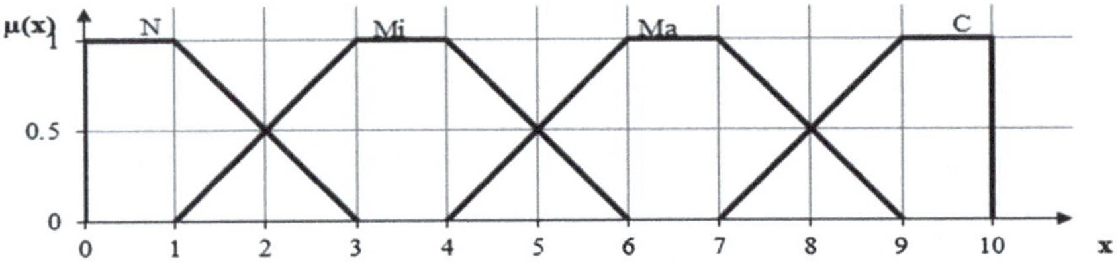

Fig. 5.5 Linguistic assessment scales and membership degrees for risk level

5.5 Results

The parameters in the proposed method were used to analyze the COVID-19 transmission risk of three main mining environments with different production methods and working conditions. The linguistic assessments made by the experts were calculated by the fuzzy inference mechanism and the rule base in the proposed method was used to analyze the risk of transmission. The evaluations are activity-based and analyzed separately. The linguistic responses of the assessments and the results are given in Table 5.7 for the open-pit mine, Table 5.8 for the underground metal mine, and Table 5.9 for the underground coal mine.

Table 5.6 Rule base

Number of workers	Working distance	Type of working environment																								
		VL					L					M					H					VH				
		Working time in cooperation					Working time in cooperation					Working time in cooperation					Working time in cooperation					Working time in cooperation				
		VL	L	M	H	VH	VL	L	M	H	VH	VL	L	M	H	VH	VL	L	M	H	VH	VL	L	M	H	VH
VL	VL	N	N	N	N	N	N	N	N	N	N	N	N	N	N	N	N	N	N	N	N	N	N	N	N	N
	L	N	N	N	N	N	N	N	N	N	N	N	N	N	N	N	N	N	N	N	N	N	N	Mi	Mi	Mi
	M	N	N	N	N	N	N	N	N	N	N	N	N	N	N	N	N	N	N	N	Mi	Mi	Mi	Mi	Mi	Mi
	H	N	N	N	N	N	N	N	N	N	Mi	N	Mi	Mi	Mi	Mi	Mi	Mi	Mi	Mi	Ma	Mi	Mi	Mi	Mi	Ma
	VH	N	N	N	N	Mi	N	Mi	Mi	Mi	Mi	Mi	Mi	Mi	Mi	Mi	Mi	Mi	Ma	Ma	Ma	Mi	Mi	Mi	Ma	Ma
L	VL	N	N	N	N	N	N	N	N	N	N	N	N	N	N	N	N	N	Mi	Mi	Mi	N	N	Mi	Mi	Mi
	L	N	N	N	N	N	N	N	N	N	N	N	Mi	Mi	Mi	Mi	N	Mi	Mi	Mi	Mi	Mi	Mi	Ma	Ma	Ma
	M	N	N	N	N	N	N	N	N	N	N	Mi	Mi	Mi	Mi	Mi	Mi	Mi	Mi	Mi	Ma	Mi	Mi	Ma	Ma	Ma
	H	N	N	N	N	N	N	N	N	N	Mi	Mi	Mi	Mi	Mi	Ma	Ma	Mi	Mi	Ma	Ma	Mi	Mi	Ma	C	C
	VH	N	N	N	N	Mi	N	N	N	Mi	Mi	Mi	Mi	Mi	Ma	Ma	Ma	Ma	Ma	Ma	Ma	Ma	Ma	C	C	C
M	VL	N	N	N	N	Mi	N	N	N	N	N	N	N	N	N	Mi	N	N	N	Mi	Mi	N	N	Mi	Mi	Ma
	L	N	N	N	N	Mi	N	N	N	Mi	Mi	N	N	Mi	Mi	Mi	Mi	Mi	Mi	Ma	Ma	Mi	Ma	Ma	Ma	Ma
	M	N	N	N	Mi	Mi	N	Mi	Mi	Mi	Mi	Mi	Mi	Ma	Ma	Mi	Mi	Mi	Ma	Ma	Ma	Mi	Ma	Ma	Ma	C
	H	N	N	N	Mi	Mi	N	Mi	Mi	Mi	Ma	Mi	Ma	Ma	Ma	Ma	Mi	Ma	Ma	Ma	Ma	Ma	Ma	C	C	C
	VH	N	N	N	Mi	Mi	Mi	Mi	Mi	Ma	Ma	Ma	Ma	Ma	Ma	Ma	Ma	Ma	Ma	C	C	Ma	C	C	C	C
H	VL	N	N	N	Mi	Mi	N	Mi	Mi	Mi	Mi	N	N	Mi	Mi	Mi	Mi	Mi	Mi	Mi	Ma	Mi	Mi	Mi	Mi	Ma
	L	N	N	Mi	Mi	Mi	N	Mi	Ma	Ma	Ma	N	Ma	Mi	Ma	Mi	Mi	Mi	Ma	Ma	Ma	Mi	Mi	Ma	Ma	Ma
	M	N	N	Mi	Mi	Mi	Mi	Ma	Ma	Ma	Ma	Ma	Ma	Ma	Ma	Ma	Mi	Ma	Ma	Ma	C	Mi	Ma	Ma	C	C
	H	N	N	Mi	Mi	Ma	Ma	Ma	Ma	Ma	Ma	Ma	Ma	Ma	Ma	Ma	Ma	Ma	Ma	C	C	Ma	Ma	C	C	C

(continued)

Table 5.6 (continued)

Number of workers	Working distance	Type of working environment																								
		VL					L					M					H					VH				
		Working time in cooperation					Working time in cooperation					Working time in cooperation					Working time in cooperation					Working time in cooperation				
		VL	L	M	H	VH	VL	L	M	H	VH	VL	L	M	H	VH	VL	L	M	H	VH	VL	L	M	H	VH
	VH	N	Mi	Mi	Ma	Ma	Ma	Ma	Ma	Ma	Ma	Ma	Ma	Ma	C	C	Ma	Ma	C	C	C	Ma	C	C	C	C
VH	VL	N	N	N	N	Mi	N	N	N	N	Mi	N	N	N	Mi	Mi	N	Mi	Mi	Mi	Ma	Mi	Mi	Ma	Ma	Ma
	L	N	N	N	Mi	Mi	N	N	Ma	Ma	Ma	N	N	Mi	Ma	Ma	Mi	Mi	Mi	Ma	Ma	Ma	Ma	Ma	Ma	Ma
	M	N	N	Mi	Mi	Ma	Ma	Ma	Ma	Ma	Ma	Mi	Mi	Ma	Ma	Ma	Ma	Ma	Ma	C	C	Ma	Ma	Ma	C	C
	H	N	Mi	Mi	Ma	Ma	Ma	Ma	Ma	Ma	C	Ma	Ma	C	C	C	C	C	C	C	C	C	C	C	C	C
	VH	Mi	Ma	Ma	Ma	Ma	Ma	Ma	Ma	Ma	C	Ma	Ma	C	C	C	C	C	C	C	C	C	C	C	C	C

Table 5.7 Assessment of the open-pit mine and the risk level

Activities	Number of workers	Working distance	Working environment	Working time	Risk magnitude
Drilling	L	VL	VL	VH	N
Blasting	M	VL	VL	VH	Mi
Loading	L	VL	VL	VH	N
Transportation (minerals)	VH	VL	VL	VH	Mi
Crushing and grinding	L	L	VL	VH	N
Ground support	L	M	VL	M	N
Drainage	L	VL	VL	M	N
Occupational health and safety	L	VL	VL	M	N
Electrical and mechanical activities	L	L	VL	H	N
Mineral processing	H	L	VL	VH	Mi

Table 5.8 Assessment of the underground metal mine and the risk level

Activities	Number of workers	Working distance	Working environment	Working time	Risk magnitude
Drilling	M	VL	M	VH	Mi
Blasting	H	VL	M	VH	Mi
Loading	L	VL	M	VH	N
Transportation (minerals)	M	VL	M	VH	Mi
Transportation (workers)	VH	VH	M	VL	Ma
Ground support	M	M	M	VH	Mi
Drainage	VL	VL	M	M	N
Occupational health and safety	L	VL	M	M	N
Electrical and mechanical activities	L	L	M	H	Mi
Mineral processing	H	L	M	VH	Mi

Table 5.9 Assessment of the underground coal mine and the risk level

Activities	Number of workers	Working distance	Working environment	Working time	Risk magnitude
Excavation	VH	L	M	VH	Ma
Loading	L	VL	M	VH	N
Transportation (minerals)	L	VL	M	VH	N
Transportation (workers)	VH	VH	M	VL	Ma
Drilling	L	L	M	H	Mi
Blasting	L	VL	M	L	N
Ground support	L	M	M	VH	Mi
Drainage	L	VL	M	M	N
Occupational health and safety	L	VL	M	M	N
Electrical and mechanical activities	L	L	M	H	Mi
Mineral processing	H	L	VL	VH	Mi

According to the results of the open-pit mine, if there is a COVID-19 carrier in the environment, there is no high or catastrophic transmission risk. It has been determined that the risk of transmission is low in blasting, ore transportation, and mineral processing and insignificant in all other activities.

The results of the underground metal mine show that high transmission risk was detected only if there were COVID-19 carriers in the environment during the transportation activities of the workers to the working area. Apart from these activities, it has been determined that the risk of contamination is low in drilling, blasting, and mineral processing while the risk in the rest of the activities is minor.

The results of the underground coal mine show that the high transmission risk was detected in the presence of COVID-19 carriers in the environment during the excavation and the transportation of the workers to the working area. Apart from these activities, it was determined that the risk of contamination in drilling, ground support, electrical and mechanical activities, and mineral processing is low and insignificant in all other activities.

5.6 Discussion

This study investigated the transmission risk of COVID-19 in mining activities. The Mamdani fuzzy inference system was used to characterize the mining activities and the parameters. In the open-pit mine, which is almost entirely mechanized and produced valuable minerals outdoors, the transmission risk is at an acceptable level, although more workers are needed compared to the underground metal mines. However, the high transmission risk was determined in some of the activities of underground mines.

In underground operations, personnel transportation is an activity that increases the risk of transmission even though the duration of the activity is the shortest compared to all activities. As a result of environmental conditions, the number of people, and particularly social distance, the high risk of transmission is inevitable. In a simulation study conducted in Wuhan, where the virus appeared for the first time, it was emphasized that the most important transmission factor was social distance (Prem et al. 2020). Therefore, the co-working distance and working environment are the two most important factors for the transmission in the mines. In addition, according to the results of the current study, the transmission risk is lower with mechanized production that enables to comply with social distancing rules and decrease co-working time. Therefore, underground coal mine has more risk of COVID-19 transmission than the other two mines.

In a study carried out between January–March 2020, when the virus was most active for the 30 states in China, the number of cases was analyzed according to meteorological conditions such as temperature and humidity. A significant difference was observed between 1 °C increase in temperature and a decrease in the number of cases. When the obtained data is interpreted, it was concluded that low temperature and low humidity may increase the risk of COVID-19 transmission, but this result does not express certainty (Liu et al. 2020). In our study, although the risk of transmission is independently evaluated from the ambient conditions, there may be quite different meteorological conditions in the mining activities. A new study should be performed by placing these parameters on a precise model.

5.7 Conclusion

The mining sector has been one of the most influential factors in reaching the level of technology and welfare of developed countries throughout history. Developed countries that use their natural resources effectively owe their economic power mainly to mining activities. Thus, these activities should continue in a sustainable manner without being affected by the pandemic. The continuation of these activities depends on a good assessment of the risk of transmission and taking the necessary measures. This study indicated the risk of COVID-19 transmission in different mining activities. The methodology is based on a method of fuzzy inference system that characterizes the risk by means of several parameters that depend on the working environment. The results show that the transmission risk is at acceptable levels in mining environments where social distance can be achieved (open working environment and mechanized production). However, the high risk of transmission can be seen in environments where production is carried out in closed areas with an unmechanized system. The findings emphasize the necessity of mechanized production and social distance. The proposed approach can be useful for decision-makers and researchers who assess COVID-19 transmission risk. Future research will be focused on placing the new parameters, such as temperature, relative humidity, wind speed, and type of surface, on a new precise model.

References

Ahamed S, Samad M (2020) Information mining for covid-19 research from a large volume of scientific literature. arXiv:2004.02085

Baker SR, Bloom N, Davis SJ, Terry SJ (2020) Covid-induced economic uncertainty. Nat Bureau Econ Res 0898–2937

Camastra F et al (2015) A fuzzy decision system for genetically modified plant environmental risk assessment using Mamdani inference. Expert Syst Appl 42(3):1710–1716

Chen LD (2020) Effects of ambient temperature and humidity on droplet lifetime–a perspective of exhalation sneeze droplets with COVID-19 virus transmission. Int J Hyg Environ Health 113568

Cinar U, Cebi S (2019) A hybrid risk assessment method for mining sector based on QFD, fuzzy logic and AHP. In: International conference on ıntelligent and fuzzy systems, pp 1198–1207. Springer

Cinar U, Ugurlu OF, Cebi S (2020) Design of noise-canceling earmuffs with quality function deployment. In: Customer oriented product design, pp 23–34. Springer

Debnath J, Biswas A, Sivan P, Sen KN, Sahu S (2016) Fuzzy inference model for assessing occupational risks in construction sites. Int J Ind Ergon 55:114–128

Ebrat M, Ghodsi R (2014) Construction project risk assessment by using adaptive-network-based fuzzy inference system: an empirical study. KSCE J Civ Eng 18(5):1213–1227

Fernandes N (2020) Economic effects of coronavirus outbreak (COVID-19) on the world economy. Available at SSRN 3557504

Guimarães ACF, Lapa CMF (2007) Fuzzy inference to risk assessment on nuclear engineering systems. Appl Soft Comput 7(1):17–28

Hassen HB, Elaoud A, Salah NB, Masmoudi A (2020) A SIR-poisson model for COVID-19: evolution and transmission ınference in the maghreb central regions. Arab J Sci Eng 1–10

Ilbahar E, Karaşan A, Cebi S, Kahraman C (2018) A novel approach to risk assessment for occupational health and safety using Pythagorean fuzzy AHP & fuzzy inference system. Saf Sci 103:124–136

Jamshidi A, Yazdani-Chamzini A, Yakhchali SH, Khaleghi S (2013) Developing a new fuzzy inference system for pipeline risk assessment. J Loss Prev Process Ind 26(1):197–208

Karasan A, Ilbahar E, Cebi S, Kahraman C (2018) A new risk assessment approach: safety and critical effect analysis (SCEA) and its extension with Pythagorean fuzzy sets. Saf Sci 108:173–187

Kim H, Kim K, Kim H (2016) Vision-based object-centric safety assessment using fuzzy inference: monitoring struck-by accidents with moving objects. J Comput Civ Eng 30(4):04015075

Laing (2019) The economic impact of the coronavirus 2019 (Covid-2019): implications for the mining industry. Extract Ind Soc

Liu J et al (2020) Impact of meteorological factors on the COVID-19 transmission: a multi-city study in China. Sci Total Environ 138513

Mamdani EH (1977) Application of fuzzy logic to approximate reasoning using linguistic synthesis. IEEE Trans Comput 12:1182–1191

Mamdani E, Assilian S (1999) An experiment in linguistic synthesis with a fuzzy logic controller. Int J Hum Comput Stud 51(2):135–147

Prem K et al (2020) The effect of control strategies to reduce social mixing on outcomes of the COVID-19 epidemic in Wuhan, China: a modelling study. Lancet Pub Health

Ratnayake RC (2014) Application of a fuzzy inference system for functional failure risk rank estimation: RBM of rotating equipment and instrumentation. J Loss Prev Process Ind 29:216–224

Rezaee MJ, Yousefi S, Eshkevari M, Valipour M, Saberi M (2020) Risk analysis of health, safety and environment in chemical industry integrating linguistic FMEA, fuzzy inference system and fuzzy DEA. Stoch Env Res Risk Assess 34(1):201–218

Rosario DK, Mutz YS, Bernardes PC, Conte-Junior CA (2020) Relationship between COVID-19 and weather: case study in a tropical country. Int J Hyg Environ Health 113587

Ugurlu OF, Kumral M (2019) Optimization of drill bit replacement time in open-cast coal mines. Int J Coal Sci Technol 6(3):399–407

Wang Y-M, Elhag TM (2008) An adaptive neuro-fuzzy inference system for bridge risk assessment. Expert Syst Appl 34(4):3099–3106

WHO (2019) Coronavirus disease 2019 (COVID-19) strategic preparedness and response plan: accelerating readiness in the eastern mediterranean region: February 2020. World Health Organization. Regional Office for the Eastern Mediterranean

WHO (2020a) Getting your workplace ready for COVID-19: how COVID-19 spreads, 19 March 2020. World Health Organization

WHO (2020b) Overview of public health and social measures in the context of COVID-19: interim guidance, 18 May 2020. World Health Organization

Chapter 6
Simulation Output Analysis for Risk Assessment and Mitigation

David Fernando Muñoz

Abstract In this Chapter we discuss the main techniques for output analysis of simulation experiments, with an emphasis on the estimation of performance measures for the assessment and mitigation of risk. After a review of the main concepts related to stochastic simulation and risk assessment, we discuss techniques available for estimating expectations, nonlinear functions of expectations, quantiles and M estimators, we consider transient simulation as well as steady-state simulation. We discuss methodologies for both point estimation and the assessment of the accuracy of point estimators for performance measures, and the application of output analysis techniques is illustrated through examples related to risk assessment and mitigation.

Keywords Stochastic simulation · Risk measurement · Risk modeling · Simulation applications

6.1 Introduction

According to Aven (2016), the area of risk assessment and management has evolved considerably since its beginnings in the 1970s, and there have been developed a wide variety of methods and applications in most societal sectors. As evidence of this evolution, we can observe the variety of research groups of the Society for Risk Analysis, among which we can mention: Dose Response, Ecological Risk Assessment, Emerging Nanoscale Materials, Engineering and Infrastructure, Exposure Assessment, Microbial Risk Analysis, Occupational Health and Safety, Risk Policy and Law, and Security and Defense.

Aven (2016) also mentions that the area of risk assessment and management has two fundamental tasks: (i) to use risk assessments and management to study and treat the risk caused by the execution of specific activities (for example, the operation of an offshore facility or investment), and (ii) conduct research and development (in

D. F. Muñoz (✉)
Department of Industrial and Operations Engineering, Instituto Tecnológico Autónomo de México, Rio Hondo 1, 14200 Mexico City, Mexico
e-mail: davidm@itam.mx

J. Ren (ed.), *Multi-Criteria Decision Analysis for Risk Assessment and Management*, Industrial Ecology and Environmental Management 1,
https://doi.org/10.1007/978-3-030-78152-1_6

general) on risk, developing concepts, theories, frameworks, approaches, principles, methods and models to understand, evaluate, characterize, communicate and (in a broad sense) manage and mitigate risk.

Parallel to the development of the area of risk assessment and mitigation, concepts, techniques and available tools (software) have been developed for systems simulation and, in particular, for stochastic simulation, which is the type of simulation that allows us to include uncertainty and risk components in a model. In practice, a model for risk management can become complex, in the sense that we cannot obtain analytical expressions for the risk measures that are relevant to the problem under study and, in such circumstances, stochastic simulation has particular relevance for the estimation (from the output of simulation experiments) of the risk measures to be mitigated.

The objective of this Chapter is to present a review of the techniques that have been proposed to analyze the output of simulation experiments, in order to estimate performance measures that are important to conduct a risk assessment and mitigation study, when a simulation model is used to imitate the evolution of a system.

The Chapter is organized as follows. After this introduction, we present a brief literature review on the relevant applications of systems simulation for risk assessment and mitigation. In the next section, we present an overview of the necessary concepts and tools available to conduct simulation experiments. The following section discusses the most important techniques for estimating risk measures in transient simulations, including the estimation of expectations, variances and risk measures based on quantiles and M estimators. In this Section, we also present a Bayesian framework to incorporate parameter uncertainty in the process of estimating risk measures. Finally, the last section discusses the techniques available to estimate risk measures in steady-state simulations, considering again the estimation of expectations, variances, and quantile-based risk measures. In the last section we also discuss the initial transient problem and how can it be mitigated.

6.2 Literature Review

In this section we present a brief review on the main literature related to simulation applications that have been successfully applied for risk assessment and mitigation in different areas. The literature on the applications of risk assessment and mitigation is abundant, and this is only a very brief review on the applications of simulation in this area, for a more detailed review on the applications of risk assessment and mitigation, the reader is referred to Aven (2016).

According to Aven (2016), an important step in the process of making informed decisions for risk management corresponds to risk assessment, which consists of the analysis of the knowledge base to have an understanding about the risks and the uncertainties related to the case under study. As explained in Aven (2012), although it is true that the criteria for evaluating risks are usually based on the estimation of expected values (e.g., the cost of a negative event) or probabilities (of a negative event), we can find arguments for the use of other measures for risk assessment.

For example, in the area of finance, risk measures have been proposed based not only on the estimation of expected losses, but also on quantile-based measures, such as the Value at Risk (VaR) or the Conditional Value at Risk (CVar), see e.g., Natarajan et al. (2009). Because of these reasons, in addition to the techniques for estimating expectations and probabilities from the output of simulation experiments, in this Chapter we will also deal with the estimation of other risk measures, such as the variance and risk measures based on quantiles and M estimators, recognizing that some other measures for risk management and mitigation could be proposed in addition to the ones discussed in this Chapter.

Stochastic simulation has been widely used for risk assessment in various areas, for example, in supply chain management, where risk measures are mainly related to shortages, the occurrence of catastrophes and the costs incurred (see, e.g., Wu and Olson 2008; Wu et al. 2012; Chen et al. 2013; Hamdi et al. 2018; Oliveira et al. 2019; and their references). Stochastic simulation has also been used extensively in the areas related to production planning to design products with high reliability, for example, for water distribution (see, e.g., Wagner et al. 1988; Ostfeld et al. 2002), for the design of integrated circuits (see, e.g., Hu 1992; Wang et al. 2007; Li et al. 2008), or for the design of highly reliable products (see, e.g., Heidelberger 1995; Juneja and Shahabuddin 2006; Bucklew 2013). One area of production planning where stochastic simulation is particularly important for risk mitigation is operations scheduling, where the achievement of programs that meet delivery dates is very important (see, e.g., Pegden 2017; Smith et al. 2019).

In areas related to health care, stochastic simulation experiments have also been successfully conducted, for example, to design spaces for medical care with a low risk of experiencing long waiting times (see, e.g., Fone et al. 2003), to improve the understanding and mitigation of epidemics (see, e.g., Salathe et al. 2012), to make economic evaluations of diseases and their treatments (see, e.g., Cooper et al. 2006). A more complete review of the applications of simulation for health care can be found in Mielczarek and Uziałko-Mydlikowska (2012).

Simulation has been successfully applied in the areas of waste treatment and energy recovery (see, e.g., Ren et al. 2010; Ren 2018; Liang et al. 2020; Yang et al. 2020), and to mitigate the risk of the occurrence of landslides (see, e.g., Dai et al. 2002; Fell et al. 2005), or to quantify the resilience of power systems (see, e.g., Pantelli et al. 2017) or urban infrastructure (see e.g., Ouyang and Duenas-Osorio 2012).

6.3 Systems Simulation for Risk Assessment and Mitigation

In this section we present an overview of the main concepts related to systems simulation for risk assessment and mitigation as well as the software available for building models for stochastic simulation.

6.3.1 Systems Simulation

The term system is used in various disciplines to identify the elements and dynamics of a phenomenon that is intended to be understood, analyzed and/or designed from the point of view of the corresponding discipline. According to Schmidt and Taylor (1970), a system is a collection of entities that interact to achieve a goal. For example, in Industrial Engineering we study industrial systems (supply chains, service centers, manufacturing plants, etc.) that consist of raw materials, human resources and capital, organized to efficiently produce and distribute manufactures and/or services. In the same way, systems can be studied in Economics from the point of view of the welfare of the agents involved in the economic phenomenon and, similarly, each discipline study systems from its analytical perspective.

Without a doubt, humanity has studied systems from very ancient times. Initially, an attempt was made to understand natural systems through experimentation with the real system. The search for knowledge led to the development, first of physical models of systems (prototypes, scale models, etc.) that allowed them to carry out controlled experiments, and later, theories and mathematical models that could explain and predict the behavior of systems, both existing ones and those that were developed. A physical model is an imitation, generally simpler, of a real system, whose experimentation (under controlled conditions) allows us to study the behavior of the system in a natural way, as it would happen with the real system. A mathematical model, on the other hand, represents the system to be studied by means of mathematical relationships; therefore, by experimenting with it, we can predict the behavior of the relevant variables of the system and imagine the main behavior of the system, even if it is not physically reproduced.

One of the purposes of a mathematical model is to predict the behavior of one or more characteristics of the system (known as response variables) based on other variables (called control variables). A mathematical model in which, through a set of equations, the response variables are expressed as a (explicit) function of the control variables is very convenient to predict the behavior of a system, and we say that the model has an analytical solution when this set of equations exists.

However, when we want to study a system in great detail, we must consider variables whose relationships are not easy to solve to find an analytical solution. Nonetheless, the model can still be useful to analyze the system, since for this purpose numerical methods have been developed. Given particular values for the control variables, numerical methods allow us to calculate, by using a computer, the value of the response variables.

Among the numerical methods used to study a system (see Fig. 6.1), simulation has the fundamental characteristic that the model tries to imitate the behavior of the system under study, in order to calculate, with the help of a computer, the value of the system's response variables. For the purposes of this Chapter, we will recognize by simulation the *computer imitation of the behavior of a system, using a (mathematical) model to explain its relevant characteristics, in order to numerically evaluate the performance measures of the system.*

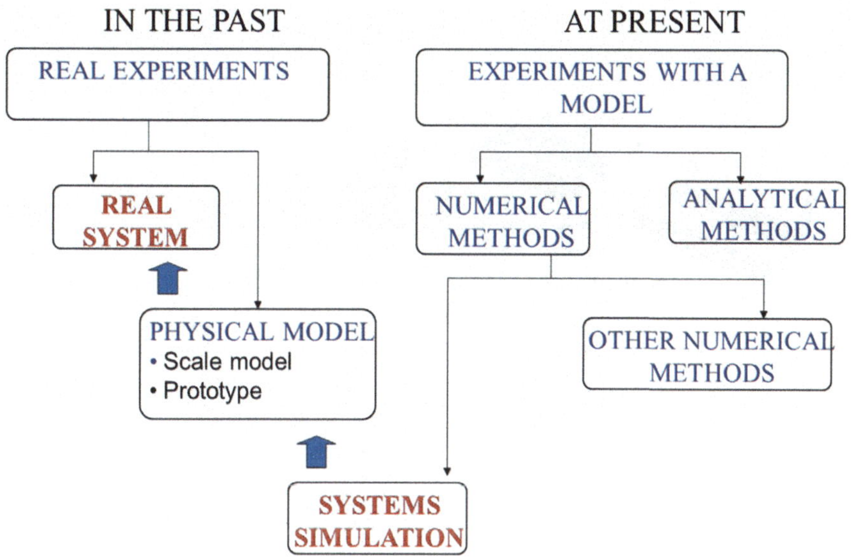

Fig. 6.1 Simulation as a tool to study a system

Because in a simulation we try to imitate a system, simulation experiments can be animated, that is, according to the results that the model is providing, the evolution of the system over time can be illustrated by developing an interface animation (see Fig. 6.2). Therefore, a simulation experiment can be viewed as a virtual experiment, with many advantages over real experimentation, among which we can mention:

i. The behavior of a system can be simulated without affecting it, and therefore simulation experimentation is less expensive than real experimentation.
ii. The relevant cost in simulation experimentation is that of model development and validation, and many experiments can be simulated with relatively low additional costs.
iii. Simulation is a very useful tool for the design of prototypes and/or new systems, because we can simulate the behavior of systems that do not exist.
iv. Real experimentation can hinder the normal operation of the system, and in the case of systems in which people interact, it is known that they can change their normal behavior when they feel observed.

A main feature of many of the tools available to develop simulation experiments is that they allow modeling the uncertainty on some characteristics of the system under study; which is very convenient for evaluating the risk and uncertainty that is present in many real-life situations. For example, if we wish to study customer service at a telephone exchange, it would be unrealistic to assume that we know precisely when customer calls will occur; therefore, it would be convenient for the simulation model to generate the occurrences following some pattern that incorporates the

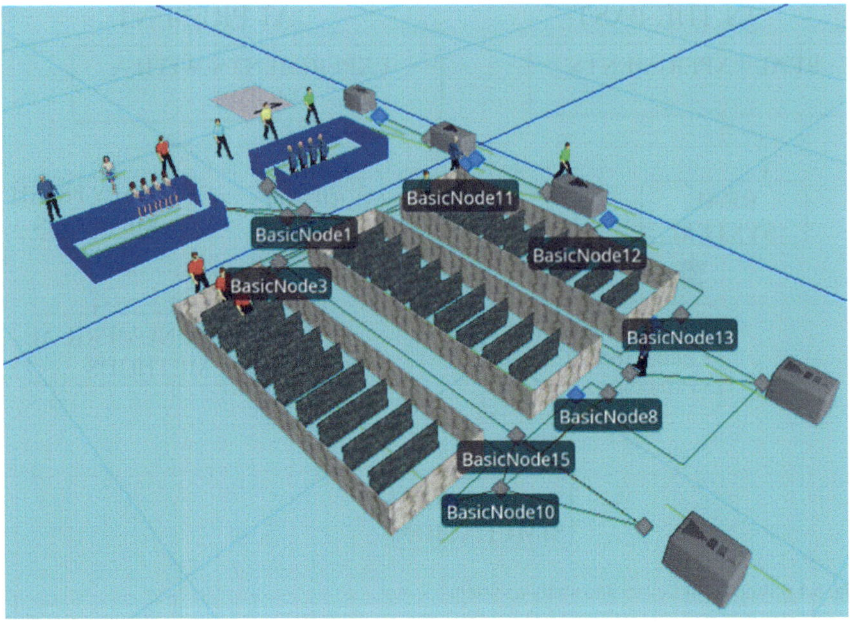

Fig. 6.2 A view of an animation in Simio for an emergency room

uncertainty about these occurrences. As is well known, uncertainty can be quantified and incorporated into a model by using the concept of probability. Customer calls to the telephone exchange can be modeled assuming that the time between successive calls is a random variable that follows some probability distribution; for example, assuming they follow an exponential distribution with a five-minute average between calls. The elements of a simulation model that are generated using some probability distribution are called random components of the model; and a model that has random components is called a stochastic model. As is expected, in order to simulate stochastic models, we must use mechanisms to generate (simulate) random variates (observations) from the probability distributions that are incorporated into the model. Simulations that use random variate generators are also known as Monte Carlo simulations.

6.3.2 Steps Suggested for a Study Using Stochastic Simulation

The development and analysis of simulation experiments based on stochastic models is performed in a similar way as real experimentation, although we must be careful for both modeling the random components of the model and analyzing the output of the simulation experiments, since it must be taken into account that, because

Fig. 6.3 Steps suggested for a study using stochastic simulation

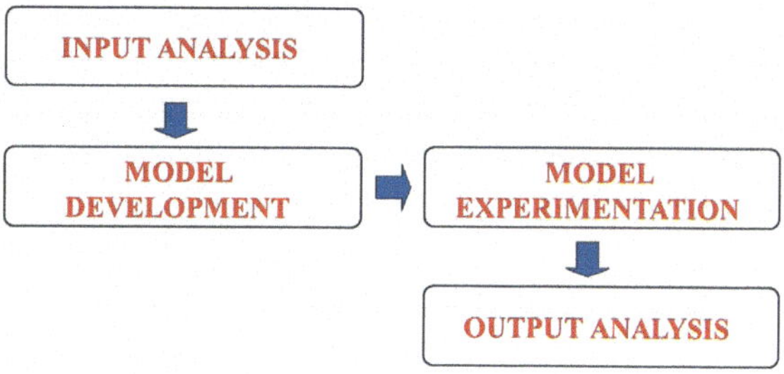

the model has random components, the result of a particular simulation run is also random. Therefore, no conclusions can be made based on a single replication (run) of the experiment, and the experiment must be replicated to draw statistically significant conclusions. It is worth mentioning that, when real experiments are analyzed, conclusions cannot be drawn based on a single observation of the phenomenon, since no matter how much control one has over all the variables in the experiment, there is a variability in the results from an observation to another; This is how experimentation with stochastic models also resembles or "simulates" what happens with real experiments.

As illustrated in Fig. 6.3, there are four fundamental steps to studying a system using stochastic simulation:

i. Input analysis. Consists of determining the components of the model and their parameters. In this step, we determine which components are deterministic (there is no uncertainty) and which are random. Furthermore, in the latter case, the probability distributions (and the value of their parameters) must be specified, this is usually achieved by considering data from the real system, forecasting models and even subjective information in the absence of quantitative data (for further details see, e.g., Muñoz 2014).
ii. Model development. Consists of developing the computer code for the basic simulation model that represents the system to be analyzed. After selecting the tools to be used to develop the model, verification and consistency tests are carried out, both to verify that no programming errors have been made and to verify that the developed simulation model can, indeed, reproduce the characteristics of the real system (for a more detailed explanation, see Smith et al. 2019).
iii. Model experimentation. Consists of designing the experiments and recording the experimental results that will allow us to make valid conclusions related to the system under study. For example, when designing the distribution of a vaccine, the values of the parameters (e.g., temperature) must be established to reduce the risk of deterioration at a reasonable cost.
iv. Output analysis: Consists of analyzing the information obtained from experimentation with the model to establish statistically valid conclusions about the system under study. In this step, care must be taken to apply the appropriate statistical techniques, which will be discussed in this Chapter.

It is worth mentioning that a professional simulation study can consume important resources (time and money), depending on its scope and the stakeholders involved in the study, so the project must be carefully planned. The readers interested in the details for a good planning of a simulation study are referred to Chap. 1 of Smith et al. (2019).

6.3.3 Available Software for Stochastic Simulation

The beginning of stochastic simulation dates to the 1950s, when the first Monte Carlo methods were implemented in computers to solve problems of integral calculus. The first applications of simulation, during the 1960s, were developed mainly in research centers, by highly specialized personnel and, generally, using the FORTRAN language on relatively expensive and bulky computers. In the 1970s and 1980s, the potential of simulation as a systems' analysis tool was recognized, simulation languages were developed, and the main models used in simulation were identified.

As a first classification of simulation models, we distinguish between *static models* and *dynamic models*, whose fundamental difference is that in static models time does not appear as a model variable or, in other words, a static model is used to study a system at a particular moment of time, while using a dynamic model we study the evolution of a system over time. The most used dynamic models in simulation are systems dynamics and discrete-event simulation. In simulation of systems dynamics, the state changes over time are modeled by means of ordinary differential equations, and numerical methods for integration are applied (see, e.g., Logan 2012). Among the most popular languages for this type of simulation are DYNAMO and CSSL. Systems dynamics simulation is also known as continuous simulation.

In discrete-event simulation, the state of the system does not change continuously over time; to be more precise, the state of the system changes only in discrete instants of time and due to the occurrence of some event. For example, the inventory of products in a warehouse changes only when a shipment of products arrives (increases), or when a delivery is made (decreases), so the natural simulation of inventories in a warehouse corresponds to discrete-event simulation. This type of simulation has found important applications in risk assessment and mitigation, as it is used to analyze inventory policies, task scheduling, supply chain performance, vehicle scheduling policies and, in general, to evaluate the performance of different policies to manage engineering systems and, in particular, production and distribution of manufactures and services. Among the most popular discrete-event simulation languages we can mention: SIMAN, GPSS, SIMSCRIPT and SLAM.

Simulation languages facilitated the application of simulation as a design tool, but they still had the disadvantage that their use required some learning time. For this reason, in the 1990s, the advantages of personal computers and window-based operating systems were considered, and simulation languages were introduced in packages with graphical and interactive environments that facilitate the development and animation of simulation applications. In software for discrete-event simulation, such as SIMIO, ARENA, PROMODEL, WITNESS or QUEST, the old commands of programming languages are presented as modules that are integrated into a graphical programming environment, allowing the animation of the simulation models (see Fig. 6.2). Currently, languages and packages that integrate different simulation models (continuous and discrete-event) have been developed. In the near future, as the communication between different operating systems is facilitated, it is expected that the simulators will be integrated into specialized applications by industrial sector, or by groups of companies, which will allow their use in an easy and natural way to solve problems of design or scheduling and control of tasks in real time.

At present, simulation models can be developed in a variety of software (see Fig. 6.4), from spreadsheets using an add-in or a library developed in a general purpose language (e.g., C++), to special purpose software with capabilities for 3D animation, or using languages such as C++, Python or R, which have libraries for random variate generation. Since in this Chapter we will use simple models to illustrate the application of methods for output analysis, we developed a small C++ library whose Windows installer is available in the software section of http://ciep.itam.mx/~davidm under the heading *Chapter Output Analysis.* The examples developed for this Chapter can be implemented in Linux or Mac OS from this code, although to produce the results that we report in this Chapter we preferred to export the procedures to a.dll library, and then call them from a spreadsheet.

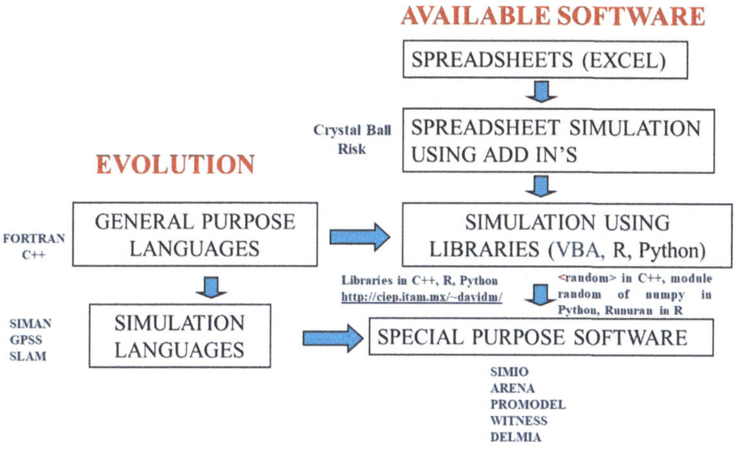

Fig. 6.4 Available software for discrete-event stochastic simulation and its evolution

6.4 Output Analysis for Transient Simulations

In general, the output of a stochastic (dynamic) simulation can be regarded as a stochastic process $\{Y(s) : t \geq 0\}$, where $Y(s)$ is a random vector (of arbitrary dimension d) representing the state of the simulation at time $s \geq 0$. The term *transient simulation* applies to a dynamic simulation that has a well-defined termination time, so that the output of a transient simulation can be viewed as a stochastic process $\{Y(s) : 0 \leq s \leq T\}$, where T is a stopping time (may be deterministic), see, e.g., Chung (2000) for a definition of stopping time. Note that this notation includes the case of a discrete-time output $X_0, X_1, \ldots$ if we assume that $Y(s) = X_{\lfloor s \rfloor}$, where $\lfloor s \rfloor$ denotes the integer part of s.

A performance variable W_1 in transient simulation is a function that depends on the simulation output up to time T, i.e., $W_1 = f(Y(s), 0 \leq s \leq T)$, and a performance measure is a parameter defined by the cumulative distribution function (c.d.f.) of W_1, i.e., a performance measure in a transient simulation can be denoted as a functional $T(F)$, where F is the c.d.f. of a performance variable W_1, i.e., $F(w) = P[W_1 \leq w]$, $w \in \mathbb{R}$. The standard methodology that is used to estimate a performance measure in transient simulation is the method of replications, that consists on running the simulation model to produce m replications $W_1, W_2, \ldots, W_m$ that can be regarded as independent and identically distributed (i.i.d.) random variables. This notation is convenient, because a natural point estimator for a performance measure $T(F)$ in this setting is just the estimator $T(F_m)$, where F_m denotes the empirical distribution of $W_1, W_2, \ldots, W_m$, defined by

$$F_m(w) = \frac{1}{m} \sum_{i=1}^{m} I(W_i \leq w), \tag{6.1}$$

for $w \in \mathbb{R}$ (see, e.g., Serfling 2009).

In this Section we will present methodologies to produce simulation-based point estimators and the corresponding accuracy assessment for the most important risk parameters $T(F)$. We will start by introducing a simple inventory example to illustrate the application of our simulation output analysis techniques.

6.4.1 A Simple Inventory Model

The simple model that we introduce below is based on the model proposed in Muñoz and Muñoz (2011) to forecast the demand of items with sporadic demand, and is inspired in the ideas of Kalchsmidth et al. (2006), where they suggest the use of forecasting techniques that take into account not only the time series, but also the structure of the process that generates the demand (non-systematic variability). In what follows we will refer to this model as Model 1.

Let us suppose that a seller uses a (Q, R) policy (see, e.g., Nahmias 2013) to order the supply of a certain product, i.e., when the inventory level reaches the reorder point (R), a quantity Q is ordered. If an order is placed at time $t = 0$ and L is the delay, then the demand for the product during the delay is

$$W_1 = \begin{cases} \sum_{i=1}^{N(L)} U_i, & N(L) > 0, \\ 0, & \text{otherwise}, \end{cases} \tag{6.2}$$

where, for $t \geq 0$, $N(t)$ is the number of clients that arrived up to time t, and for $i = 1, 2, \ldots$, U_i is the demand for client i. We assume that $U_1, U_2, \ldots$ are i.i.d. random variables and are also independent of the stochastic process $\{N(t) : t \geq 0\}$. In order to obtain analytical results for some of our performance measures, we also assume that $\{N(t) : t \geq 0\}$ is a Poisson process with rate Θ_0.

Under our assumptions, we can define the following parameters that represent important properties for the policy (Q, R) (for details on the derivation of the analytical expressions see Muñoz et al. 2013).

The expected demand is an important measure to forecast demand W_1, and is defined by

$$\mu_W = E[W_1] = \Theta_0 L \mu_U, \tag{6.3}$$

where $E[X]$ denotes the expected value of a random variable X, and $\mu_U = E[U_1]$.

The variance of demand W_1 is an important measure for the magnitude of the uncertainty on the forecast of demand W_1, and is defined by

$$\sigma_W^2 = E\left[W_1^2\right] - E[W_1]^2 = \Theta_0 L\left(\mu_U^2 + \sigma_U^2\right), \tag{6.4}$$

where $\mu_U^2 = E[U_1]^2, \sigma_U^2 = E\left[U_1^2\right] - \mu_U^2$.

Given a value R for the reorder point, an important risk measure is the probability of no stockout (called type-1 service level), and is defined by

$$\alpha_1(R) = P[W_1 \leq R] = E[I(W_1 \leq R)] \tag{6.5}$$

where, for any event A, $I(A)$ denotes the indicator random variable that takes a value of 1 if event A occurs and zero otherwise.

Given a value $0 < \alpha < 1$, the type-1 reorder point is a value for the reorder point that provides an approximate type-1 service level of α, and is defined by

$$r_1(\alpha) = \inf\{R \geq 0 : \alpha_1(R) \geq \alpha\}, \tag{6.6}$$

where $\alpha_1(R)$ is defined in (6.5).

Similarly, given a value R for the reorder point, another important risk measure is the proportion of demand that is met from stock (called type-2 service level or *fill rate*), and is defined by

$$\alpha_2(R) = 1 - \frac{E[(W_1 - R)I(W_1 > R)]}{Q}, \tag{6.7}$$

and given a value $0 < \alpha < 1$, the type-2 reorder point is a value for the reorder point that provides an approximate type-2 service level of α, and is defined by

$$r_2(\alpha) = \inf\{R \geq 0 : \alpha_2(R) \geq \alpha\}, \tag{6.8}$$

where $\alpha_2(R)$ is defined in (6.7).

In the next sections, we show how to estimate the measures of risk defined by Eqs. (6.3) through (6.6) from the output of a simulation, and remark that Model 1, as defined in (6.1), is a very simple model just to verify (using simulation) that we are proposing valid estimation procedures that may be applied to a complex model, for which we would not have analytical solutions.

6.4.2 Properties of a Good Estimator

In order to discuss the main properties that a good simulation-based estimator must satisfy, we use the concept of weak converge of random variables. We say that a sequence of random variables $X_1, X_2, \ldots$ converge weakly to a random variable X (and denote $X_m \Rightarrow X$, as $m \to \infty$), if $\lim_{m\to\infty} F_{X_m}(x) = F_X(x)$, at any point x where F is continuous, where F_{X_m} and F_X denote the c.d.f. of X_m and X, respectively (see, e.g., Chung 2000). Note that X can also be a constant (M), on which case X is simply the random variable that takes the value of M with probability 1.

A first property that a good estimator must satisfy is consistency. We say that the estimator $T(F_m)$, where F_m is defined in (6.1) is consistent if

$$T(F_m) \Rightarrow T(F), \tag{6.9}$$

as $m \to \infty$. Note that consistency means that the estimator $T(F_m)$ approaches the parameter $T(F)$ as the sample size m increases, and this is a required property since we do not want the estimator to converge to a different value (or not to converge at all).

In order to assess the accuracy of a consistent estimator $T(F_m)$, we usually verify if a Central Limit Theorem (CLT) in the form of

$$\frac{\sqrt{m}(T(F_m) - T(F))}{\sigma_T} \Rightarrow N(0, 1), \tag{6.10}$$

as $m \to \infty$ is satisfied, on which case we may also look for a consistent estimator $\hat{\sigma}^2(m)$ for the asymptotic variance σ_T^2, so that if follows from (6.10) and standard properties of weak convergence (see, e.g., Serfling 2009) that

$$\frac{\sqrt{m}(T(F_m) - T(F))}{\hat{\sigma}(m)} \Rightarrow N(0, 1), \tag{6.11}$$

as $m \to \infty$, where $N(0, 1)$ denotes a random variable distributed as Normal with mean 0 and variance 1. It is worth mentioning that the CLT of (6.10) implies consistency of $T(F_m)$, as defined in (6.9).

A CLT in the form of (6.11) is sufficient to assess the accuracy of the point estimator $T(F_m)$, since (6.11) implies that

$$\lim_{m\to\infty} P\left[|T(F_m) - T(F)| \leq z_\beta \frac{\hat{\sigma}(m)}{\sqrt{m}}\right] = 1 - \beta,$$

where z_β denotes the $(1 - \beta/2)$-quantile of a $N(0, 1)$, which is sufficient to establish an asymptotically valid $(1 - \beta)100\%$ confidence interval (CI) for $T(F)$ with halfwidth

$$HW_T = \frac{t_{(m-1,\beta)}\hat{\sigma}(m)}{\sqrt{m}}, \tag{6.12}$$

where $t_{(m-1,\beta)}$ denotes the $(1 - \beta/2)$-quantile of a Student-t distribution with $(m - 1)$ degrees of freedom. A halfwidth in the form of (6.12) is the typical measure used in simulation software to assess the accuracy of $T(F_m)$ for the estimation of a parameter $T(F)$. Note that we are using a Student-t distribution to have a wider CI when the value of m is small, and the CI is still asymptotically valid since $t_{(m-1,\beta)} \to z_\beta$, as $m \to \infty$. Note also from (6.12) that, to lower the value of a halfwidth (i.e., improve the accuracy of $T(F_m)$), we need to increase the sample size m, so that the halfwidth will be reduced approximately by half if we multiply the sample size m by 4.

6.4.3 Estimation of Expected Values

For the estimation of the expected value $T(F) = E[W_1]$, the point estimator $T(F_m)$ becomes the sample average

$$\bar{W}(m) = \frac{\sum_{i=1}^{m} W_i}{m}, \tag{6.13}$$

and it is well-known from the classical CLT that the CLT (6.10) is satisfied for $T(F_m) = \bar{W}(m)$, and $\sigma_T^2 = E\left[W_1^2\right] - E[W_1]^2$. Moreover, since $W_1, W_2, \ldots, W_m$ are i.i.d., it is well known that a consistent (and unbiased) estimator for σ_T^2 is

$$S_W^2(m) = \frac{\sum_{i=1}^{m} \left(W_i - \bar{W}(m)\right)^2}{m-1}, \tag{6.14}$$

so that, it follows from (6.11) that an asymptotically valid $(1-\beta)100\%$ halfwidth for the expected value $\mu_W = E[W_1]$ is given by

$$HW_{\mu_W} = \frac{t_{(m-1,\beta)} S_W(m)}{\sqrt{m}}, \tag{6.15}$$

where $S_W(m)$ is defined in (6.14).

Thus, Eq. (6.13) can be used to compute a point estimator for the expected value in a transient simulation, and Eq. (6.15) allows us to compute an assessment of the accuracy of the point estimator. Note that Eqs. (6.13) and (6.15) can be applied not only to the estimation of the expected demand (6.3) in Model 1 but also for the estimation of a type-1 service level defined in (6.5) or a type-2 service level defined in (6.7), since the service levels are also expectations, to be more precise, we can take $W_{1i} = I[W_i \leq R]$ for parameter (6.5) and $W_{2i} = 1 - (W_i - R)I[W_i > R]/Q$ for parameter (6.7), $i = 1, \ldots, m$. A C++ code for the estimation of these risk measures using simulation output was compiled to produce a library and below we report numerical examples using this code.

6.4.4 Estimation of the Variance and Nonlinear Functions of Expectations

Suppose that we have a multivariate output $X_i = (X_{i1}, X_{i2}, \ldots, X_{id})$, where X_i corresponds to the i-th independent replication of a transient simulation and, given a nonlinear function $g : \mathbb{R}^d \to \mathbb{R}$, we are interested in the estimation of $T(F) = g\left(\mu_{X_1}\right)$, where $\mu_{X_1} = (\mu_1, \mu_2, \ldots, \mu_d)$, and $\mu_j = E\left[X_{1j}\right]$, $j = 1, 2, \ldots, d$.

If g is differentiable at μ_X and $\sigma_g^2 = \nabla g(\mu_X)^T V \nabla g(\mu_X) < \infty$, it can be shown that (see, e.g., Muñoz and Glynn 1997)

$$\frac{\sqrt{m}\left(g\left(\bar{X}(m)\right) - g(\mu_X)\right)}{\sigma_g} \Rightarrow N(0,1), \tag{6.16}$$

where V is the covariance matrix of $X_1 = (X_{11}, X_{12}, \ldots X_{1d})$. Furthermore, a consistent (although typically biased) estimator for σ_g^2 is

$$\hat{\sigma}_g^2(m) = \nabla g(\bar{X}(m))^T \hat{V}(m) \nabla g(\bar{X}(m)), \tag{6.17}$$

where $\bar{X}(m) = \left(\bar{X}_1(m), \bar{X}_2(m), \ldots \bar{X}_d(m)\right)$, $\bar{X}_j(m) = \left(\sum_{i=1}^{m} X_{ij}\right)/m$, $j = 1, 2, \ldots, d$, and $\hat{V}(m)$ is a consistent estimator for V (it can be easily obtained from $X_1, X_2, \ldots, X_m$). Therefore, it follows from (6.11) to (6.16) that, an asymptotically valid $(1-\beta)100\%$ halfwidth for $g(\mu_X)$ is given by

$$HW_{g(\mu_x)} = \frac{t_{(m-1,\beta)}\hat{\sigma}_g(m)}{\sqrt{m}}, \tag{6.18}$$

where $\hat{\sigma}_g(m)$ is defined in (6.17).

For the case of risk measure (6.4) in Model 1 (demand variance) we can take $d = 2$, $X_i = (W_i, W_i^2), i = 1, 2, \ldots, m$, and $g(x_1, x_2) = x_2 - x_1^2$, so that $g(\mu_{X_1}) = E[W_1^2] - E[W_1]^2 = \sigma_W^2$, $S_W^2(m)$ is a consistent estimator for σ_W^2, $\nabla g(x_1, x_2) = (-2x_1, 1)^T$, and

$$\begin{aligned}\hat{\sigma}^2_{\sigma^2_W}(m) =& \nabla g(\bar{X}(m))^T \hat{V}(m) \nabla g(\bar{X}(m)) \\ =& 4\bar{W}(m)^2 S_W^2(m) - 4\bar{W}(m)S_{12}(m) + S_2^2(m),\end{aligned} \tag{6.19}$$

where $\bar{W}(m)$ is defined in (6.13), $S_W^2(m)$ is defined in (6.14), and

$$S_2^2(m) = \frac{\sum_{i=1}^m (W_i^2 - \bar{W}_2(m))^2}{m-1}, \; S_{12}(m) = \frac{\sum_{i=1}^m (W_i - \bar{W}(m))(W_i^2 - \bar{W}_2(m))}{m-1},$$

$$\bar{W}_2(m) = \frac{\sum_{i=1}^m W_i^2}{m}.$$

Then, it follows from (6.18) that an asymptotically valid $(1-\beta)100\%$ halfwidth for the variance $\sigma_W^2 = E[W_1^2] - E[W_1]^2$ is given by

$$HW_{\sigma^2_W} = \frac{t_{(m-1.\beta)}\hat{\sigma}_{\sigma^2_W}(m)}{\sqrt{m}}, \tag{6.20}$$

where $\hat{\sigma}_{\sigma^2_W}(m)$ is defined in (6.19).

So far, we have provided consistent estimators and asymptotically valid halfwidths for the performance measures of Model 1 defined in Eqs. (6.3)–(6.5) and (6.7). Note that, since demand is discrete, for risk measures (6.6) and (6.8) it suffices to provide consistent estimators because a consistent estimator will take the value of the parameter for a large-enough number of replications m. If we consider the corresponding $T(F_m)$ estimators for the functionals $T(F)$ defined in (6.6) and (6.8), respectively, we obtain the estimators

$$\hat{r}_i(\alpha) = \inf\{R \geq 0 : \alpha_i(R) \geq \alpha\}, \tag{6.21}$$

$i = 1, 2$, respectively, where

$$\hat{\alpha}_1(R) = \frac{\sum_{i=1}^m I(W_i \leq R)}{m}, \; \hat{\alpha}_2(R) = 1 - \frac{\sum_{i=1}^m [(W_i - R)I(W_i \leq R)]}{mQ}.$$

Table 6.1 Point estimators and halfwidths for risk measures in a simulation of Model 1 ($m = 1000$)

Parameter	Point estimator	Halfwidth	Lower bound	Upper bound
Mean	90.7740	0.9378	89.8362	91.7118
Variance	324.4854	24.8561	299.6293	349.3415
Reorder T1	114			
Reorder T2	94			
T1 service level	0.9000	0.0156	0.8844	0.9156
T2 service level	0.9524	0.0044	0.9481	0.9568

Note that, for R given, $\hat{\alpha}_1(R)$ and $\hat{\alpha}_2(R)$ are the corresponding consistent estimators for type-1 and type-2 service levels, respectively. As we are going to see later in this Chapter, the estimators defined in (6.21) are not only consistent but they also satisfy a CLT in the form of (6.10).

To illustrate our results, we run some experiments using our interface in Excel and the procedures implemented in the spreadsheet corresponding to Model 1. In all our experiments we considered a lead time of $L = 15$, an arrival rate of $\Theta_0 = 2$, and $P[U_1 = i] = 0.2, i = 1, 2, \ldots, 5$. Nominal service levels (α) for type-1 and type-2 reorder points are 0.90 and 0.95, respectively, the reorder quantity (Q) was set in 120, and the confidence level for halfwidths is $1 - \beta = 0.9$. Using Eqs. (6.3) and (6.4) we can verify that the expected demand and its variance are $\mu_W = 90$ and $\sigma_W^2 = 330$, respectively, which allows us to verify consistency of our estimation procedures using simulation. In Table 6.1 we present a typical output of one simulation experiment for $m = 1000$ independent replications. Note that the last two columns correspond to the point estimator minus and plus the halfwidth, respectively. As we see from Table 6.1, the asymptotic CI for both mean and variance covered the true values, and service levels for point estimators of type-1 and type-2 reorder points are slightly larger than the nominal values (0.9 and 0.95, respectively), which is typical in a discrete-demand setting.

In order to empirically verify consistency of our point estimators, we replicated the experiment reported in Table 6.1 by multiplying de value of of m by four each time ($m = 4000$, 16000, 64000 and 256000). In all the experiments the asymptotic CI for both mean and variance covered the true values and, as expected, the halfwidths were lowered by a factor closed to half each time we multiply the value of m by 4 (see Fig. 6.5). Note that, in Fig. 6.5 we are reporting the relative errors (halfwidth divided by the point estimator) in order to report the values in the same graph. Although we are not reporting halfwidths for the reorder points estimations, we mention that the estimated type-1 reorder points were 114 in all cases, except for $m = 4000$, and the estimated type-2 reorder points were 94 for $m = 1000$, 4000, and 93 for $m =$ 16000, 64000, 256000, confirming that these estimations converge to 114 and 93, respectively (the true values).

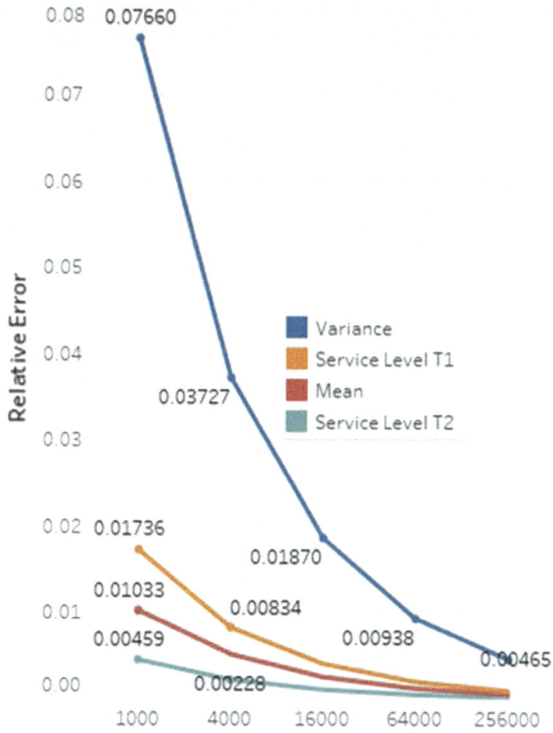

Fig. 6.5 Relative errors in simulations of Model 1 for different risk measures and values of m

6.4.5 Parameter Uncertainty and Bayesian Estimation

Most of the proposed techniques to compute service levels and reorder points assume that, as in Model 1, the parameters of the model that is used for demand forecasting are known with certainty (see, e.g., Nahmias 2013). However, in practice, parameters are estimated from available information (data and/or expert judgment), and there exists a certain degree of uncertainty in the value of these parameters. We now describe a Bayesian framework that allows us to incorporate parameter uncertainty that is induced from the estimation procedure.

To describe a Bayesian framework for inventory management (as in Muñoz et al. 2013), we need to slightly change our notation, we now assume that $W_1 = g(Y(s), 0 \leq s \leq T; \Theta)$ is the demand for an item, where $\{Y(s), s \geq 0; \Theta\}$ is a stochastic process (possibly multivariate), T is a stopping that represents the planning horizon, and Θ is a (random) vector of parameters with state-space S. Information on Θ is available through a vector of (real) observations $x \in \mathbb{R}^d$ that has a likelihood function $L(x|\theta)$. If $p(\theta)$ is a prior density function for the vector of parameters Θ, according to Bayes' theorem the posterior (conditional on data x) density function of Θ becomes

For $j = 1$ to m

1. Generate (independently) a value θ_j from $p(\theta|x)$.
2. Run a simulation experiment with $\Theta = \theta_j$ to obtain an independent replication

$$W_j = g\big(Y_j(s), 0 \le s \le T; \Theta = \theta_j\big)$$

End Loop

Compute point estimators and halfwidths for inventory management as usual.

Fig. 6.6 Posterior Sampling algorithm

$$p(\theta|x) = \frac{p(\theta)L(x|\theta)}{\int_S p(\theta)L(x|\theta)d\theta}, \tag{6.22}$$

for $x \in \mathbb{R}^d$ and $\theta \in S$.

The measures of risk for inventory management under this framework are conditional on the value of observations $X = x$. For example, the appropriate expression for the expected demand is $\mu_W = E[W_1|X = x]$, and similarly for the other measures of risk defined in Eqs. (6.4) through (6.8). Fortunately, we do not need to change too much our estimations procedures since, if we have a valid procedure to generate random variates from $p(\theta|x)$, we can applied the Posterior Sampling (PS) algorithm illustrated in Fig. 6.6, where we indicate that we first simulate a value for parameter Θ (from the posterior distribution), and then we run the simulation experiment with a fixed value for the parameter. We say that we compute point estimators and halfwidths as usual to indicate that these estimators are computed as in the case of Model 1, we only need to consider that all performance measures are now conditional on the event $X = x$, i.e., any performance measure $T(F)$ reported for Model 1 has now F defined by

$$F(w) = P[W_1 \le w|X = x], w \in \mathbb{R}$$

Next, we introduce a more realistic model (called Model 2) to illustrate the application of the PS algorithm to incorporate parameter uncertainty in our simulation experiments.

Let us consider a demand defined by (6.2) (as in Model 1). However, we assume that the arrival rate Θ_0 and the probabilities $\Theta_j = P[U_1 = j], j = 1\ldots q$ for an individual demand are uncertain (although the maximum demand q is known). Set $\Theta = \big(\Theta_0, \Theta_1, \ldots \Theta_{q-1}\big)$, and suppose that the information on Θ comes from a random sample of n clients: $v = (v_1, \ldots, v_n)$, $u = (u_1, \ldots, u_n)$, where v_i is the interarrival time between client i and client $i-1$, and u_i is the demand for client i, $i = 1, \ldots, n$. Under the assumptions of Model 1, the likelihood function is a product of exponentials and multinomial densities:

$$L(v|\theta) = \theta_0^n e^{-\theta_0 \sum_{i=1}^{n} v_i} \left(1 - \sum_{j=1}^{q-1} \theta_j\right)^{c_q} \prod_{j=1}^{q-1} \theta_j^{c_j}, \tag{6.23}$$

for $\theta_0 > 0$, $\theta_j \geq 0$, $j \geq 1$, $\sum_{j=1}^{q-1} \theta_j \leq 1$, where $\theta = (\theta_0, \theta_1, \ldots \theta_{q-1})$, and $c_j = \sum_{i=1}^{n} I(u_i = j)$ is the number of clients that ordered j items, $j = 1, \ldots, q$. As is known, Jeffrey's (objective) prior for the exponential distribution is $p(\theta_0) = \theta_0^{-1}$ (see, e.g., Bernardo and Smith 2000), and for the multinomial distribution is the Dirichlet distribution (Berger and Bernardo 1992):

$$p(\theta_1, \ldots, \theta_{q-1}) = \frac{\left(1 - \sum_{j=1}^{q-1} \theta_{1j}\right)^{-1/2} \prod_{j=1}^{q-1} \theta_{1j}^{-1/2}}{B(1/2, \ldots, 1/2)},$$

where $B(a_1, \ldots, a_q) = \left(\prod_{i=1}^{q} \Gamma(a_i)\right)\left(\Gamma\left(\sum_{i=1}^{q} a_i\right)\right)^{-1}$, so that, if we adopt an objective point of view, it follows from (6.22) to (6.23) that the posterior distribution is the product of gamma and Dirichlet:

$$p(\theta|x) = \frac{\theta_0^{n-1} \left(\sum_{i=1}^{n} v_i\right)^n e^{-\theta_0 \sum_{i=1}^{n} v_i}}{(n-1)!} \frac{\left(1 - \sum_{j=1}^{q-1} \theta_j\right)^{c_q - 1/2} \prod_{j=1}^{q-1} \theta_j^{c_j - 1/2}}{B(c_1 + 1/2, \ldots, c_q + 1/2)}, \tag{6.24}$$

where $x = (v, u)$.

As we would expect, the variance of demand W_1 (conditional on $X = x$) under a Bayesian approach tends to be larger than under the approach of fixing the value of parameters, moreover, it follows from the PS algorithm that

$$\begin{aligned} V[W] &= E[W^2] - E[W]^2 = E[E[W^2|\Theta] - \mu(\Theta)^2] + E[\mu(\Theta)^2] - E[\mu(\Theta)]^2 \\ &= \sigma_S^2 + \sigma_P^2, \end{aligned} \tag{6.25}$$

where $\mu(\Theta) = E[W|\Theta]$, $\sigma_S^2 = E[\sigma^2(\Theta)]$ is called the *stochastic variance*, $\sigma_P^2 = V[\mu(\Theta)]$ is called the *parametric variance* and, to simplify the notation, the operator V means "variance conditional on $X = x$" and E means "expectation conditional on $X = x$". Under the approach of fixing the value of parameters we have $\sigma_P^2 = 0$, and a larger variance under a Bayesian approach is justified by the fact that the uncertainty on the value of parameters is considered.

It is worth mentioning that the PS algorithm of Fig. 6.6 allows us to estimate $\sigma_S^2 + \sigma_P^2$. If we want to estimate σ_P^2 alone, we must have more than one replication in step 2. However, as shown in Muñoz (2017), more than one replication in step 2 is not computationally efficient for the estimation of the expected demand (and perhaps for the estimation of other risk measures).

For Model 2 we can estimate the same risk measures as for Model 1, as shown below, see Muñoz et al. (2013) for hints on how to derive the analytical expressions.

The expected demand is defined by

$$\mu_W = E[W_1|X = x] = E[T\Theta_0]\sum_{j=1}^{q} jp_j, \tag{6.26}$$

where $p_j = \frac{\alpha_j}{\alpha_0}, \alpha_j = c_j + \frac{1}{2}, \alpha_0 = \sum_{j=1}^{q} \alpha_j, E[T\Theta_0] = \frac{Tn}{\sum_{j=1}^{n} v_i}$.

The variance of demand W_1 is defined by

$$\sigma_W^2 = E\left[W_1^2|X = x\right] - E[W_1|X = x]^2 = \sigma_P^2 + \sigma_S^2, \tag{6.27}$$

where $\sigma_P^2 = \frac{E\left[T^2\Theta_0^2\right]}{(\alpha_0+1)}\sum_{j=1}^{q} j^2 p_j + \frac{E[T\Theta_0]^2[(\alpha_0/n)-1]}{(\alpha_0+1)}\left(\sum_{j=1}^{q} jp_j\right)^2, \sigma_S^2 = E[T\Theta_0]\sum_{j=1}^{q} j^2 p_j, E\left[T^2\Theta_0^2\right] = T^2 n(1+n)/\left(\sum_{j=1}^{n} v_i\right)^2$.

Given a value R for the reorder point, the type-1 service level is defined by

$$\alpha_1(R) = P[W_1 \leq R|X = x] = E[I(W_1 \leq R)|X = x] \tag{6.28}$$

Given a value $0 < \alpha < 1$, the type-1 reorder point is defined by

$$r_1(\alpha) = \inf\{R \geq 0 : \alpha_1(R) \geq \alpha\}, \tag{6.29}$$

where $\alpha_1(R)$ is defined in (6.28).

Similarly, given a value R for the reorder point, the type-2 service level, is defined by

$$\alpha_2(R) = 1 - \frac{E[(W_1 - R)I(W_1 > R)|X = x]}{Q}, \tag{6.30}$$

and, given a value $0 < \alpha < 1$, the type-2 reorder point is defined by

$$r_2(\alpha) = \inf\{R \geq 0 : \alpha_2(R) \geq \alpha\}, \tag{6.31}$$

where $\alpha_2(R)$ is defined in (6.30).

As we mentioned before, performance measures under a Bayesian approach have the same form $T(F)$ as in the traditional approach, with the only difference that $F(w) = P[w_1 \leq w|X = x], w \in \mathbb{R}$, so that all the estimation procedures for Model 1 apply to Model 2 with $W_1, \ldots, W_m$ defined in the PS algorithm of Fig. 6.6. To illustrate our results, we run some experiments using our interface in Excel and the procedures implemented in the spreadsheet corresponding to Model 2. In all our experiments we considered input data similar to the one used for Model 1, a lead time of $L = 15$, $n = 20$ clients, $\sum_{j=1}^{n} v_j = 10$, $q = 5$, and $c_i = 4, i = 1, 2, \ldots, 5$.

Nominal service levels (α) for type-1 and type-2 reorder points are 0.90 and 0.95, respectively, the reorder quantity (Q) was set in 120, and the confidence level for halfwidths is $1 - \beta = 0.9$. Using Eqs. (6.26) and (6.27) we can verify that the expected demand and the variance are $\mu_W = 90$ and $\sigma_W^2 \approx 815.4255$, respectively, which allow us to verify consistency of our estimation procedures using simulation. A typical output of one simulation experiment is as in Table 6.1 for Model 1, and we run experiments for $m = 1000$, 4000, 16000, 64000 and 256000, as we did with Model 1.

In all the experiments the asymptotic CI for both mean and variance covered the true values, except for the case of the mean and $m = 1000$ (for a 90% CI we expect this to happen in 1 out of 10 times). As expected, the halfwidths were lowered by a factor closed to half each time we multiplied the value of m by 4 (see Fig. 6.7). The estimated type-1 reorder points were 130 for m =1000, 127 for $m = 4000$, 64000, and 128 for $m = 16000$, 256000, and the estimated type-2 reorder points were 105 in all cases, except for $m = 1000$, confirming that these estimations converge to 128 and 105, respectively (the true values).

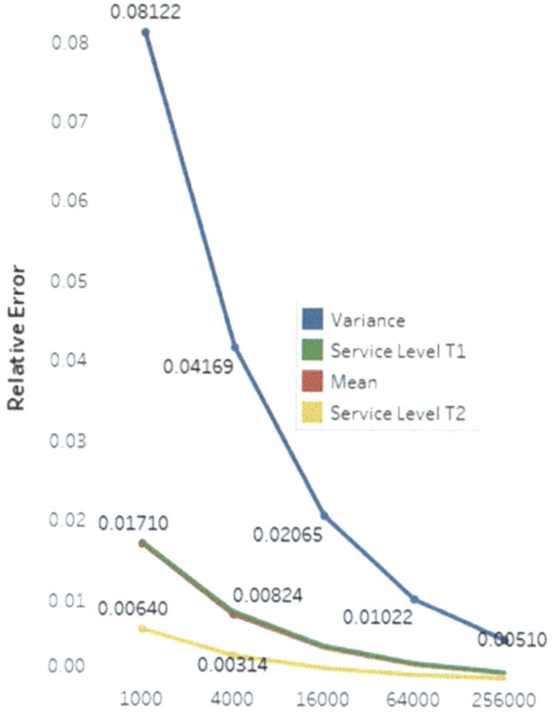

Fig. 6.7 Relative errors in simulations of Model 2 for different risk measures and values of m

6.4.6 Estimation of Risk Measures Based on Quantiles and M Estimators

We remark that in models 1 and 2 demand W_1 was discrete, and halfwidths for reorder points were not required because we know that the point estimator will hit the corresponding true value for m large enough. However, for a continuous performance variable W_1, halfwidths for the reorder points will be required in order to assess the accuracy of the point estimators. To illustrate how to compute reorder points for a continuous performance variable, we consider a Model 3 based on Model 2, with the main difference that demand U_1 for a customer is distributed as uniform on $(0, \Theta_1)$. Information on parameter $\Theta = (\Theta_0, \Theta_1)$ comes from a random sample of n clients: $v = (v_1, \ldots, v_n)$, $u = (u_1, \ldots, u_n)$, where v_i is the interarrival time of client i and u_i is the corresponding demand, $i = 1, \ldots, n$. Then, under the assumptions of Model 1, the likelihood function is a product of exponential distributions and uniforms:

$$L(v|\theta) = \theta_0^n e^{-\theta_0 \sum_{i=1}^{n} v_i} \theta_1^{-n}, \tag{6.32}$$

for $\theta_0 > 0$, $\theta_1 > M = \max\{u_1, \ldots, u_n\}$, where $\theta = (\theta_0, \theta_1)$. As is known, Jeffrey's (objective) prior for the exponential distribution is $p(\theta_0) = \theta_0^{-1}$ and for the uniform distribution is also $p(\theta_1) = \theta_1^{-1}$, for $\theta_1 > 0$ (see, e.g., Bernardo and Smith 2000). Then, an uninformative (improper) prior is $p(\theta) = \theta_0^{-1}\theta_1^{-1}$, so that, if we adopt an objective point of view, it follows from (6.22) to (6.32) that the posterior distribution is the product of gamma and Pareto:

$$p(\theta|x) = \frac{\theta_0^{n-1}\left(\sum_{i=1}^{n} v_i\right)^n e^{-\theta_0 \sum_{i=1}^{n} v_i}}{(n-1)!} n M^n \theta_1^{-(n+1)}, \tag{6.33}$$

where $x = (v, u)$, and $\theta_0 > 0$, $\theta_1 > M = \max\{u_1, \ldots, u_n\}$.

For Model 3 we can estimate the same risk measures as for Model 2, the corresponding definitions are given by Eqs. (6.26) through (6.31). However, the analytical expressions for the mean and the variance are as follows.

$$\mu_W = E[W_1|X = x] = E[T\Theta_0]\frac{\mu_1}{2}, \tag{6.34}$$

where $E[T\Theta_0] = \frac{Tn}{\sum_{j=1}^{n} v_i}$, $\mu_1 = \frac{nM}{n-1}$, and

$$\sigma_W^2 = E\left[W_1^2|X = x\right] - E[W_1|X = x]^2 = \sigma_P^2 + \sigma_S^2, \tag{6.35}$$

where $\sigma_P^2 = \frac{E[T^2\Theta_0^2]\mu_2}{4} - \frac{E[T\Theta_0]^2\mu_1^2}{4}$, $\sigma_S^2 = \frac{E[T\Theta_0]\mu_2}{3}$, $\mu_2 = \frac{nM^2}{n-2}$.

The equations we used for the computation of halfwidths in Model 1 and Model 2 are still valid for Model 3. However, we have not provided equations for the computation of halfwidths for reorder points because for discrete demand the point estimator may hit the true value for m large enough. For Model 3 the corresponding point estimators we have already proposed for reorder points are consistent. However, since the demand is now continuous, we must compute the corresponding halfwidths to assess the accuracy of the point estimators.

The type-1 reorder point $r_1(\alpha)$ is also known as the α-quantile of the probability distribution of W_1, and for quantiles we known that (see, e.g., Serfling 2009), under the assumption that the c.d.f. of W_1 is differentiable at the reorder point $r_1(\alpha)$, and $F'(r_1(\alpha)) > 0$, a CLT in the form of (6.10) is satisfied, to be more precise:

$$\frac{\sqrt{m}\left(\hat{r}_1(\alpha) - r_1(\alpha)\right)}{\sigma_{r_1}} \Rightarrow N(0, 1), \tag{6.36}$$

as $m \to \infty$, where $\hat{r}_1(\alpha)$ is defined in (6.21), and

$$\sigma_{r_1}^2 - \frac{\alpha(1-\alpha)}{[F'(r_1(\alpha))]^2}$$

As we already discussed, to obtain a halfwidth in the form of (6.12), we might want to consistently estimate $\sigma_{r_1}^2$. However, to avoid the estimation of the density $F'(r_1(\alpha))$, we will apply a result established in Sect. 2.6.3 of Serfling (2009), which shows that the following nonparametric halfwidth corresponds to an asymptotically valid $(1-\beta)100\%$ CI for $r_1(\alpha)$.

$$H_{r_1} = \left(Z_{m_{11}} + Z_{m_{12}}\right)/2, \tag{6.37}$$

where $m_{11} = \lfloor m\alpha - t_{(m-1,\beta)}[m\alpha(1-\alpha)]^{1/2} \rfloor, m_{12} = \lfloor m\alpha + t_{(m-1,\beta)}[m\alpha(1-\alpha)]^{1/2} \rfloor$ and the Z_i's are the ordered values $Z_1 \leq Z_2 \leq \ldots \leq Z_m$ obtained from $W_1, W_2, \ldots, W_m$.

A CLT in the form of (6.10) for a type-2 reorder point $r_2(\alpha)$ can be obtained from a more general result that we state as follows. For any function $\psi(x, t)$ we may associate a functional $T(F)$ as the solution t_0 of

$$\int \psi(x, t_0) dF(x) = 0,$$

and we say that T is the M-functional corresponding to ψ. As is shown in Serfling (2009), under appropriate regularity conditions for the M-functional T, a CLT in the form of (6.10) is satisfied, where the asymptotic variance has the form of

$$\sigma_T^2 = \frac{\int \psi^2(x, T(F)) dF(x)}{[\lambda'(T(F))]^2},$$

where $\lambda(t) = \int \psi(x, t) dF(x)$, for $t \in \mathbb{R}$. As shown in Muñoz et al. (2013), $r_2(\alpha)$ is an M-functional that satisfies the required conditions, and the following CLT for $r_2(\alpha)$ holds.

$$\frac{\sqrt{m}\left(\hat{r}_2(\alpha) - r_2(\alpha)\right)}{\sigma_{r_2}} \Rightarrow N(0, 1),$$

as $m \to \infty$, where $\hat{r}_2(\alpha)$ is defined in (6.21), and

$$\sigma_{r_2}^2 = \frac{\int_{r_2(\alpha)}^{\infty}(y - r_2(\alpha))^2 dF(y) - (1-\alpha)^2 Q^2}{(1 - F(r_2(\alpha)))^2}.$$

Since a consistent estimator for $\sigma_{r_2}^2$ is

$$\hat{\sigma}_{r_2}^2(m) = m(m - k_2)^{-2}\left[\sum_{j=k_2}^{m}\left(W_j - \hat{r}_2(\alpha)\right)^2 - mQ^2(1-\alpha)^2\right], \tag{6.38}$$

where $\hat{r}_2(\alpha) = Z_{k_2}$, and $Z_1 \le Z_2 \le \ldots \le Z_m$ are as in (6.37), it follows from (6.12) to (6.38) that an asymptotically valid $(1-\beta)100\%$ halfwidth for $r_2(\alpha)$ is

$$H_{r_2} = \frac{t_{(m-1,\beta)}\hat{\sigma}_{r_2}(m)}{\sqrt{m}}, \tag{6.39}$$

where $\hat{\sigma}_{r_2}^2(m)$ is defined in (6.38).

To illustrate our results, we run some experiments using our Excel interface and the procedures implemented in the spreadsheet corresponding to Model 3. In all our experiments we considered a lead time of $L = 10$, $n = 10$ clients, $\sum_{j=1}^{n} v_j = 10$, and $M = 5$. Nominal service levels (α) for type-1 and type-2 reorder points are 0.90 and 0.95, respectively, the reorder quantity (Q) was set in 35, and the confidence level for halfwidths is $1 - \beta = 0.9$. Using Eqs. (6.34) and (6.35) we can verify that the expected demand and the variance are $\mu_W \approx 15.625$ and $\sigma_W^2 \approx 137.8038$, respectively, which allow us to verify consistency of our estimation procedures using simulation. In Table 6.2 we present a typical output of one simulation experiment for $m = 1000$, and we run additional experiments for $m = 4000$, 16000, 64000 and 256000, as we did with Model 2.

Table 6.2 Point estimators and halfwidths for risk measures in a simulation of Model 3 ($m = 1000$)

Parameter	Point estimator	Halfwidth	Lower bound	Upper bound
Mean	15.8007	0.5946	15.2061	16.3953
Variance	130.4409	29.4541	100.9869	159.8950
Reorder T1	30.6632	1.6434	29.0198	32.3066
Reorder T2	25.0777	1.5291	23.5486	26.6068
T1 service level	0.9000	0.0156	0.8844	0.9156
T2 service level	0.9504	0.0080	0.9423	0.9584

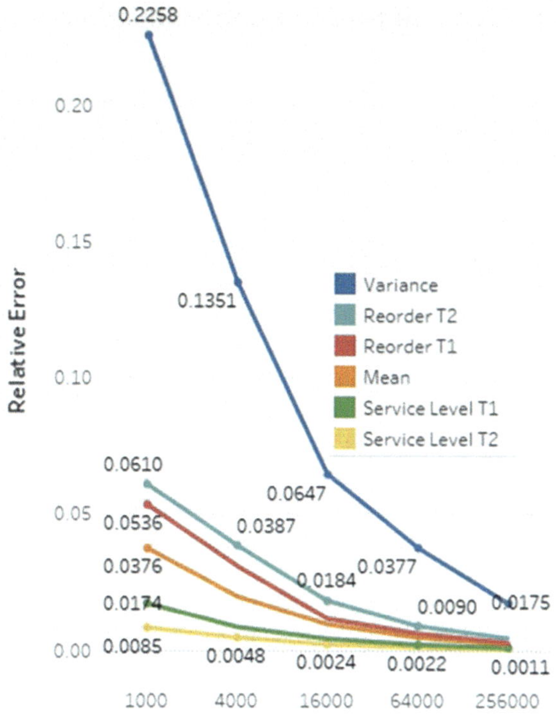

Fig. 6.8 Relative errors in simulations of Model 3 for different risk measures and values of m

In all the experiments the asymptotic CI for both mean and variance covered the true values. As expected, the halfwidths were lowered by a factor closed to half each time we multiplied the value of m by 4 (see Fig. 6.8).

The reader should notice a procedure named "Generate Sample" in the spreadsheets Model 2 and Model 3 of our interface in Excel, this procedure allows us to simulate the generation of a sample data x for the corresponding Bayesian model. We encourage the reader to try these procedures. An interesting exercise for readers with programming experience is to program the execution of independent replications of the same estimation procedure and report the empirical coverage of true parameters based on the point estimators and their corresponding halfwidths. This exercise will allow us to verify how the empirical coverage tends to the nominal confidence level $(1 - \beta)$, as m increases and the number of experiment replications is large enough (e.g., 1000).

6.5 Output Analysis for Steady-State Simulations

A second approach for running simulation-based estimation experiments is called steady-state simulation. Under this approach, we perform simulation experiments to study the long-run performance (as the time $t \rightarrow \infty$) of a system that operates without interruptions. Practical applications of steady-state simulation may arise, for example, when studying a continuous-flow production system that is assumed not to stop (e.g., production of oil or sugar) or when designing a computer and/or a communications network that never stops its service. Steady-state simulation may also be helpful as a verification tool for a simulation model of a system related to a network of queues because, under certain assumptions on the network, analytical results for steady-state performance measures may be available, so that verification tests can be applied to the simulation prototype being developed, to check if analytical results are being properly estimated (for this particular application see, e.g. Smith et al. 2019).

As one would expect, output analysis for steady-state simulations must apply to a well-defined steady-state performance measure, which usually requires that a performance variable may converge (as $t \rightarrow \infty$) to certain value that we call a steady-state performance measure. For example, for the estimation of a steady-state mean we assume that a long-run average satisfies a Law of Large Numbers (LLN) in the form of

$$\frac{\int_0^t W(s)ds}{t} \Rightarrow \mu,$$

as $t \rightarrow \infty$, for a time-persistent output $\{W(s) : s \geq 0\}$ (e.g., for the estimation of the long-run average number of clients in a queue), or

$$\frac{\sum_{i=0}^{m-1} W_i}{m} \Rightarrow \mu$$

as $m \rightarrow \infty$, for a tally output $\{W_0, W_1, \ldots\}$ (e.g., for the estimation of the long-run average clients' waiting-time in a queue). We remark that a tally output corresponds to performance variables whose observations are realized at discrete points of time. In what follows, we will use the notation for time-persistent output and, for the case of a tally output $X_0, X_1, \ldots$, we just take $W(s) = X_{\lfloor s \rfloor}$, where $\lfloor s \rfloor$ denotes the integer part of s. Note that the notion of weak convergence that we stated before, can be naturally extended for the case of a time-persistent output.

Although the steady-state output analysis literature has focused on the estimation of the steady-state mean, in this section we will present output analysis techniques for the estimation of steady-state risk measures that do not have the form of a mean, and can be applied using the same simulation experiment, we start by formulating a unified framework that allows us to define different types of steady-state risk measures.

6.5.1 Risk Measures for Steady-State Simulations

An appropriate framework for defining different types of steady-state risk measures begins by assuming that the output of the simulation converges weakly to a random variable W, i.e.,

$$W(t) \Rightarrow W, \tag{6.40}$$

as $t \to \infty$, where $\sigma_W^2 = E\left[W^2\right] - E[W] < \infty$ (the probability distribution of W is called the steady-state distribution). Then, as we did for the case of transient simulation, we can define the following steady-state parameters.

The steady-state mean is defined by

$$\mu_W = E[W]. \tag{6.41}$$

The steady-state variance is an important measure of dispersion of the steady-state distribution around its mean, and is defined by

$$\sigma_W^2 = E\left[W^2\right] - E[W]^2. \tag{6.42}$$

Given a value R, an important risk measure is the probability that the steady-state distribution does not exceed R (for the case of a service, this is called a type-1 service level), and is defined by

$$\alpha_1(R) = P[W \leq R] = E[I(W \leq R)], \tag{6.43}$$

for example, in an emergency room for a hospital, R can be 15 min, so that $\alpha_1(R)$ is the (steady state) probability that a patient will wait no more than 15 min to receive a healthcare service.

For a given number $0 < \alpha < 1$, the α-quantile of the steady-state distribution is defined by

$$r_1(\alpha) = \inf\{R \geq 0 : \alpha_1(R) \geq \alpha\}, \tag{6.44}$$

for example, $r_1(\alpha)$ can be the level at which a service is promised (e.g., a lead time to delivering a product) with a risk of $(1 - \alpha)$ of exceeding the promised level.

Note that assumption (6.40) is required to define a steady-state quantile. Furthermore, the validity of estimation methods for steady-state simulations usually require assumptions that are stronger than (6.40). For example, in the Markov chain literature, the following assumption is usually required to prove the validity of some steady-state estimation procedures.

Let us suppose that the tally output $\{W_0, W_1, \ldots\}$ is a Markov chain with state-space E and stationary distribution π. Let

$$d_k(y) \stackrel{def}{=} \sup_{A \subseteq E} |P[W_k \in A | W_0 = y] - \pi(A)|,$$

we say that $\{W_0, W_1, \ldots\}$ is geometrically ergodic if there exist $\rho < 1$ and a real-valued function f such that

$$d_k(y) \leq f(y)\rho^k, \tag{6.45}$$

for $y \in E$ and $k = 0, 1 \ldots$ (for further details, see Meyn and Tweedie 2012). Note that geometric ergodicity is basically an assumption on the rate of converge in assumption (6.40).

6.5.2 *Estimation of Performance Measures for Steady-State Simulations*

Although there has been some discussion on the use of independent replications to estimate steady-state performance measures using simulation (see, e.g., Argon and Andradóttir 2006), it is widely accepted that, unless required because of a special situation (e.g., under parallel simulation), a long run is more efficient than shorter independent simulation runs for steady-state output analysis, so that we will focus on steady-state output analysis techniques from the output $\{W(s) : 0 \leq s \leq t\}$ of a single (long) simulation. Note that, in steady-state simulation, the output can be correlated, i.e., for $s_1 \neq s_2$, $W(s_1)$ and $W(s_2)$ are no longer independent (as was the case of independent replications for transient simulation).

As in the case of transient simulation, our proposed risk measures can be expressed as a functional $T(F)$, where F is now the steady-state distribution defined in (6.40). This notation is convenient because again, a natural point estimator for $T(F)$ is $T(F_t)$, where F_t is the empirical distribution defined by

$$F_t(x) = \frac{\int_0^t I[W(s) \leq x]ds}{t}, \tag{6.46}$$

$x \in \mathbb{R}$. In particular, the following are consistent estimators under assumption (6.40).

For the steady-state mean defined in (6.41), a consistent estimator is

$$\bar{W}(t) = \int_0^t x dF_t(x), \tag{6.47}$$

For the steady-state variance defined in (6.42), a consistent estimator is

$$S_W^2(t) = \frac{t}{t-1}\left(\int_0^t x^2 dF_t(x) - \bar{W}(t)^2\right), \tag{6.48}$$

where the multiplier $t/(t-1)$ is included to provide unbiased estimation when the output corresponds to i.i.d. random variables.

For a given value R, a consistent estimator for the type-1 service level defined in (6.43) is

$$\hat{\alpha}_1(R) = F_t(R), \tag{6.49}$$

where F_t is defined in (6.46).

For a given number $0 < \alpha < 1$, an estimator for the α-quantile defined in (6.44) is

$$\hat{r}_1(\alpha) = \inf\{R \geq 0 : F_t(R) \geq \alpha\}, \tag{6.50}$$

where F_t is defined in (6.46). We point out that consistency of the point estimator for the α-quantile may require a stronger assumption than (6.40). Note that, for the case of a tally output $\{W_0, W_1, \ldots\}$ the proposed point estimators have the same form as in Eqs. (6.3) through (6.6) (proposed for transient simulation).

To assess the accuracy of the point estimators using asymptotically valid halfwidths, we may need a CLT in the form of (6.10) for each risk measure, and then we estimate the asymptotic variance σ_T^2 consistently to produce a halfwidth in the form of (6.12). However, in the steady-state case, it may not be easy to estimate consistently the asymptotic variance. Note that, under stationarity and a tally output $\{W_0, W_1, \ldots\}$, for the simplest case of the estimation of the steady-state mean, the asymptotic variance of a CLT for $\bar{W}(t)$ has the form of

$$\sigma_\mu^2 = \mathrm{Var}[W_0] + 2\sum_{i=1}^{\infty} \mathrm{COV}[W_0, W_k], \tag{6.51}$$

where $\mathrm{COV}[W_0, W_k] = E[(W_0 - \mu_W)(W_k - \mu_W)]$. Although Flegal and Jones (2010) proved consistency of the spectral variance estimator for the estimation of σ_μ^2 (for a geometrically ergodic Markov chain), Eq. (6.51) suggests that consistent estimation of σ_μ^2 may require consistent estimation of some high-order covariance terms. For this reason, we suggest the use of a methodology based on batches of observations that seems to work under more general conditions compared to consistent estimation. We will refer to this methodology as a *batching approach.*

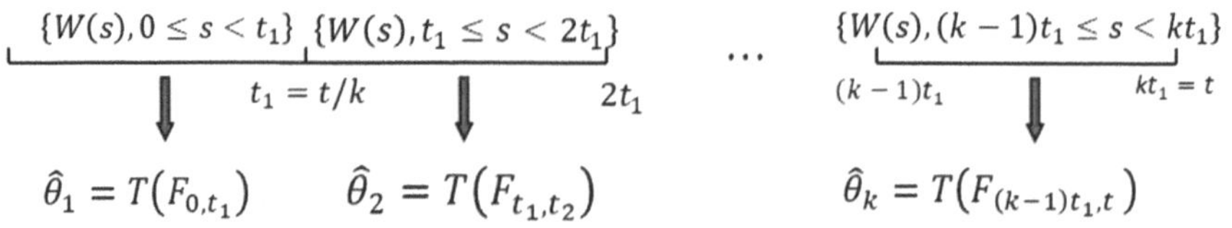

Fig. 6.9 A batching approach for a time-persistent output

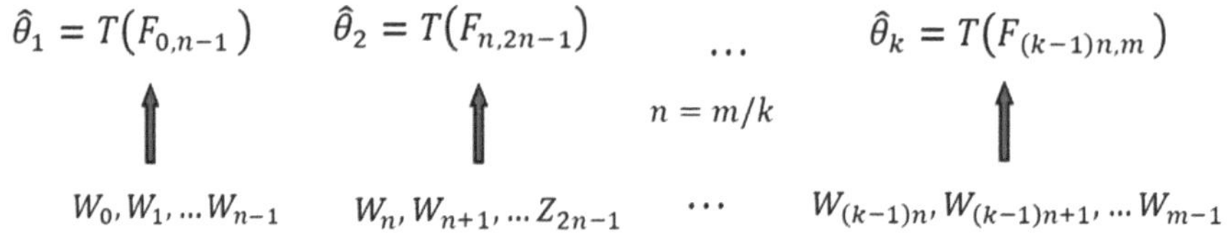

Fig. 6.10 A batching approach for a tally output

A batching approach to produce an asymptotic halfwidth for the estimation of a performance measure $T(F)$ can be implemented by following the next steps.

1. Divide the run length (t) into k batches equally spaced (in general, $5 \leq k \leq 20$ is recommended), and compute an estimator $\hat{\theta}_i = T\left(F_{t_i,t_{i+1}}\right)$ for every batch $i = 1, \ldots, k$ (see Figs. 6.9 and 6.10), where

$$F_{t_i,t_{i+1}}(x) = \frac{\int_{t_i}^{t_{i+1}} I[W(s) \leq x] ds}{t_{i+1} - t_i},$$

   is the empirical distribution for batch i, $t_i = (i-1)t/k, i = 1, \ldots, k+1$.
2. Compute the average and variance of the batch estimators:

$$\bar{\theta}(t) = \frac{\sum_{i=1}^{k} \hat{\theta}_i}{k},\ S_k^2(t) = \frac{\sum_{i=1}^{k} (\hat{\theta}_i - \bar{\theta}(t))^2}{k-1}.$$

3. Compute the halfwidth:

$$HF_T = t_{(k-1,\beta)} \frac{S_k(t)}{\sqrt{k}}. \tag{6.52}$$

Note that, for the validity of the halfwidth given in (6.52), the value of $t_{(k-1,\beta)}$ remains constant, as the run length $t \to \infty$, whereas in Eq. (6.12) for transient simulation the value of $t_{(m-1,\beta)}$ approaches the corresponding quantile of a $N(0, 1)$ distribution, as the number of replications $m \to \infty$. Sufficient conditions for the asymptotic validity of halfwidth (6.52) has been provided for the case of a steady-state mean (Glynn and Iglehart 1990), nonlinear functions of a steady-state mean

(Muñoz and Glynn 1997), steady-state quantiles (Muñoz 2010) and even for multivariate steady-state means (Muñoz and Glynn 2001). In all cases a main assumption is a slightly stronger condition than a CLT that is called a Functional Central Limit Theorem. We also mention that some other methods have been proposed for the estimation of steady-state means, e.g., regenerative method (Iglehart 1975), standardized time series (Schruben 1983), spaced batch means (Fox et al. 1991) and overlapping batch means (Meketon and Schemeiser 1984), among others. although the nonoverlapping batch means method is very easy to apply and some evidence suggest that it may produce better halfwidths (Ramírez-Nafarrate and Muñoz 2016). To illustrate our results, we run some experiments using our Excel interface and the procedures implemented in the spreadsheet corresponding to Model 4.

A useful model to test steady-state estimation procedures is the waiting time in queue of a M/M/1 queue. A fast simulation of a G/G/1 queue can be done efficiently by considering the equation

$$W_{j+1} = W_j + S_j - A_{j+1}^+, \tag{6.53}$$

for $j = 0, 1, \ldots$, where W_j is the waiting time in queue for customer j, S_j is the service time for customer j, A_j is the inter-arrival time between customer j and customer $j - 1$, and for $x \geq 0$, $x^+ = x$, otherwise $x^+ = 0$. If $S_0, S_1, \ldots$ are i.i.d. exponential random variables with rate $\mu > 0$, and $A_1, A_2, \ldots$ are i.i.d. exponential random variables with rate $\lambda > 0$, then (6.53) corresponds to a M/M/1 queue with traffic intensity $\rho = \lambda/\mu$. For a traffic intensity $\rho < 1$, the output process $W_0, W_1, \ldots$ of a simulation of an M/M/1 queue satisfies (6.40), where the c,d,f, of the steady-state distribution is

$$F_W(t) = P[W \leq t] = 1 - \rho e^{-(\mu-\lambda)t}, \tag{6.54}$$

for $t \geq 0$. From Eq. (6.54) we can obtain the following analytical results for the risk measures corresponding to a M/M/1 queue with traffic intensity $\rho < 1$.

The steady-state mean is

$$\mu_W = E[W] = \frac{\rho}{\mu(1-\rho)}. \tag{6.55}$$

The steady-state variance becomes

$$\sigma_W^2 = E\left[W^2\right] - E[W]^2 = \frac{\rho(2-\rho)}{(\mu-\lambda)^2}. \tag{6.56}$$

Given a value $c > 0$, the steady-state probability of waiting less or equal than c is

Table 6.3 Point estimators and halfwidths for risk measures of Model 4 ($m = 256000$)

Parameter	Point estimator	Halfwidth	Lower bound	Upper bound
Mean	4.0416	0.1587	3.8829	4.2003
Variance	24.8575	2.6219	22.2357	27.4794
P[W ≥ c]	0.4636	0.0059	0.4577	0.4695
α-quantile	10.4711	0.4449	10.0262	10.9160

$$P[W \leq c] = 1 - \rho e^{-c(\mu-\lambda)}. \tag{6.57}$$

For a given number $0 < \alpha < 1$, the α-quantile of the steady-distribution is

$$r_1(\alpha) = \begin{cases} 0, & \alpha \leq (1-\rho), \\ \frac{\log(\rho)-\log(1-\alpha)}{\mu-\lambda}, & \alpha > (1-\rho). \end{cases} \tag{6.58}$$

We run some experiments using our Excel interface and the procedures implemented in the spreadsheet corresponding to Model 4. In all our experiments we considered a traffic intensity of $\rho = 0.8$, an initial state of $W_0 = 0$ and $k = 10$ batches. The service level for the α-quantile was $\alpha = 0.9$, the value of c for $P[W \leq c]$ was $c = 2$ and the confidence level for halfwidths was $1 - \beta = 0.9$. Using Eqs. (6.55) through (6.58) we can verify that the true values for the risk measures are $\mu_W = 4$, $\sigma_W^2 = 24$, $P[W \leq c] \approx 0.4637$ and $r_1(\alpha) \approx 10.3972$, which allow us to verify consistency of our estimation procedures using simulation. In Table 6.3 we present a typical output of one simulation experiment for $m = 256000$, and we run additional experiments for $m = 4000$, 16000, 64000 and 1024000.

In all the experiments the asymptotic CI for all the risk measures covered the true values, except for the case of the variance and $m = 4000$ (the run length seems to be too small to be reliable). As expected, the halfwidths were lowered by a factor closed to half each time we multiplied the value of m by 4 (see Fig. 6.11), except for the case of the variance and $m = 16000$.

We end this Chapter by discussing a main problem that arises in steady-state simulation and is called the initial transient problem.

6.5.3 Warming Up

In steady-state simulation, the choice of an initial distribution that is atypical of steady-state behavior (e.g., initializing a heavy-traffic queue in the empty state) makes the first portion of the simulated observations unrepresentative of the steady-state behavior, and this is often called the *initial transient problem*. To facilitate discussion

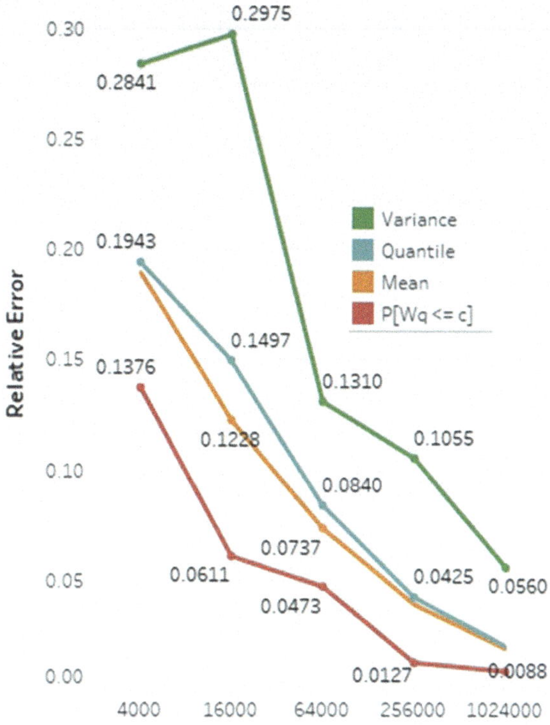

Fig. 6.11 Relative errors in a simulation of a M/M/1 queue for different risk measures and values of m

of this problem, we restrict our attention to a tally output $\{W_0, W_1, \ldots, W_m\}$ and the estimation of a steady-state mean.

Since usually we do not have a priori sense of how much bias the initial transient is generating, it is of great interest to develop procedures for determining the duration of the initial warming up (say, determined by a length $\hat{k}$) so that the first observations be discarded to consider a truncated estimator that averages over the remaining observations and is less biased than the natural estimator $\bar{W}(m)$. Initial transient deletion rules (ITDR) have been proposed to reduce the bias induced by the initial transient. However, we must be careful to choose the value of $\hat{k}$ that provides the best bias-variance tradeoff, truncating too little fails to sufficiently reduce the bias, whereas truncating too much leads to an increase in the variance of the truncated estimator.

As is well-known, the mean square error (MSE) of an estimator is, in general, the sum of the variance plus the square of the bias; and is probably because of the above mentioned bias-variance tradeoff that the MSE is the most popular criterion to judge the quality of a deletion procedure to determine $\hat{k}$. Although many ITDR have been proposed (see Robinson 2002 for a review), a very popular method (based on the MSE) is the Marginal Standard Error Rule (MSER) initially proposed in White (1997) (see Wang and Glynn 2016 for a recent analysis).

The original MSER considers a truncation point

$$\hat{k} = \underset{0 \leq k \leq m-2}{\arg\min}\, g_m(k), \tag{6.59}$$

where

$$g_m(k) \overset{def}{=} \frac{1}{(m-k)^2} \sum_{i=k}^{m-1} \left(W_i - \bar{W}(m,k)\right)^2,\ \bar{W}(m,k) \overset{def}{=} \frac{1}{m-k} \sum_{i=k}^{m-1} W_i$$

and, to avoid a zero MSE (that may happen with positive probability for discrete random variables), White et al. (2000) propose taking the arg min in (6.59) over the range $k = 0, \ldots, \lfloor m/2 \rfloor$.

Some variants (see Pasupathy and Schmeiser 2010) of the MSER rule are:

1. The MSER-LLM rule that identifies the truncation point as the first local minimizer of $g_m(k)$, and
2. MSER-LLM2 rule that identifies the truncation point as the first local minimizer among local minimizers of $g_m(k)$.

We now present some simulation experiments to illustrate how the MSER ITDR can help us to improve the quality of the steady-state mean estimation in the presence of an initial transient. We run experiments using our Excel interface and the procedures implemented in the spreadsheet corresponding to Model 4. In all our experiments we considered a traffic intensity of $\rho = 0.8$, an initial state of $W_0 = 50$ and $k = 10$ batches. The true value for the expected waiting time in queue is $\mu_W = 4$, which is far from our initial state $W_0 = 50$. Unlike previous experiments, we replicated each estimation procedures $r = 1000$ times, in order to estimate the MSE by

$$\text{MSE Est.} = \frac{\sum_{i=1}^{r} \left(\hat{\theta}_i - \mu_W\right)^2}{r},$$

where $\hat{\theta}_i$ is the estimator obtained in replication i for the corresponding ITDR. After applying an ITDR (i.e., deleting the corresponding first $\hat{k}$ values), the remaining observations were batched into $k = 10$ batches, to compute a halfwidth according to (6.52), with a confidence level of $(1 - \beta) = 0.9$. These halfwidths allowed us to compute the empirical coverage, i.e., the fraction of times the asymptotic CI covered the expected value μ_W.

Table 6.4 Results from 1000 replications of simulations of a M/M/1 queue for different ITDR and values of *m*

m	Rule	$\hat{k}$ Avg.	MSE Est.	Coverage	HW Avg.
4000	No deletion	0.00	2.6963	0.965	2.64930
4000	MSER	394.54	0.5791	0.727	0.97277
4000	MSER-LLM	159.06	0.5723	0.833	1.10906
4000	MSER-LLM2	376.20	0.7031	0.809	1.14711
16000	No deletion	0.00	0.2456	0.958	0.86501
16000	MSER	512.84	0.1278	0.841	0.57308
16000	MSER-LLM	160.26	0.1194	0.884	0.60238
16000	MSER-LLM2	464.18	0.1225	0.878	0.60525
64000	No deletion	0.00	0.0373	0.927	0.34419
64000	MSER	638.91	0.0289	0.898	0.30006
64000	MSER-LLM	157.18	0.0293	0.912	0.30685
64000	MSER-LLM2	449.41	0.0293	0.915	0.30718
256000	No deletion	0.00	0.0079	0.914	0.15990
256000	MSER	715.01	0.0074	0.899	0.15362
256000	MSER-LLM	160.28	0.0073	0.906	0.15461
256000	MSER-LLM2	462.06	0.0073	0.906	0.15496
1024000	No deletion	0.00	0.0020	0.903	0.07978
1024000	MSER	576.13	0.0020	0.891	0.07888
1024000	MSER-LLM	161.17	0.0020	0.893	0.07901
1024000	MSER-LLM2	455.52	0.0020	0.893	0.07901

We replicated the set of experiments for $m = 4000$, 16000, 64000, 256000 and 1024000 and, in each experiment, we applied 4 ITDR: no deletion, MSER, MSER-LLM and MSER-LLM2. We summarize the results of our experiments in Table 6.4 and Fig. 6.12.

As we can see from Table 6.4 to Fig. 6.12, the initial transient in our average waiting time estimation was more significant for a smaller simulation length m. Note how the coverages and halfwidths of all four ITDR tend to be the same for the largest simulation runs. In general, the no deletion rule provided the largest halfwidth average, and it also provided some over-coverage (larger than the nominal 0.9). MSER-LLM provided good coverages and reasonable halfwidths, with the smallest truncation point $\hat{k}$ (among the MSER rules).

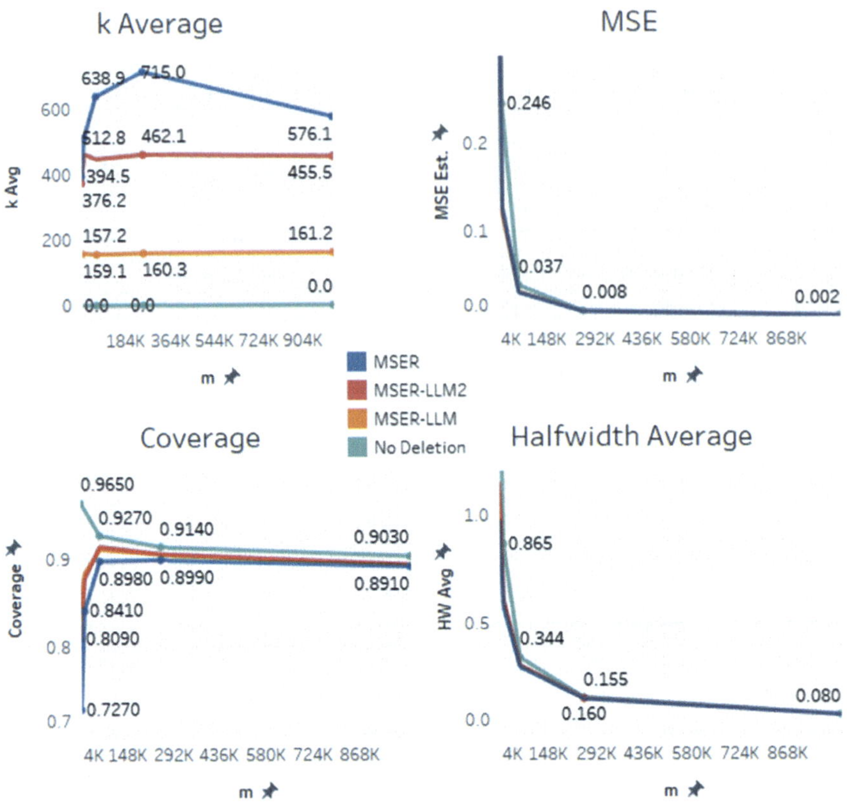

Fig. 6.12 Performance of 4 ITDR for different values of m

References

Argon NT, Andradóttir S (2006) Replicated batch means for steady-state simulations. Naval Res Logist 53(6):508–524

Aven T (2012) The risk concept—historical and recent development trends. Reliab Eng Syst Saf 99:33–44

Aven T (2016) Risk assessment and risk management: review of recent advances on their foundation. Eur J Oper Res 256:1–13

Berger JO, Bernardo JM (1992) Ordered group reference priors with application to the multinomial problem. Biometrika 79(1):25–37

Bernardo JM, Smith AFM (2000) Bayesian theory. Wiley

Bucklew J (2013) Introduction to rare event simulation. Springer

Chen X, Ong YS, Tan PS, Zhang N, Li Z (2013) Agent-based modeling and simulation for supply chain risk management-a survey of the state-of-the-art. In: 2013 IEEE international conference on systems, man, and cybernetics. IEEE, pp 1294–1299

Chung KL (2000) A course in probability theory. Elsevier

Cooper K, Brailsford SC, Davies R, Raftery J (2006) A review of health care models for coronary heart disease interventions. Health Care Manag Sci 9:311–324

Dai FC, Lee CF, Ngai YY (2002) Landslide risk assessment and management: an overview. Eng Geol 64(1):65–87

Fell R, Ho KK, Lacasse S, Leroi E (2005) A framework for landslide risk assessment and management. In: Landslide risk management. CRC Press, pp 3–25

Flegal JM, Jones GL (2010) Batch means and spectral variance estimators in Markov Chain Monte Carlo. Anna Stat 38(2):1034–1070

Fone D, Hollinghurst S, Temple M, Round A, Lester N, Weightman A et al (2003) Systematic review of the use and value of computer simulation modelling in population health and health care delivery. J Publ Health 25(4):325–335

Fox BL, Goldsman D, Swain JJ (1991) Spaced batch means. Oper Res Lett 10(5):255–263

Glynn PW, Iglehart DL (1990) Simulation output analysis using standardized time series. Math Oper Res 15(1):1–16

Hamdi F, Ghorbel A, Masmoudi F, Dupont L (2018) Optimization of a supply portfolio in the context of supply chain risk management: literature review. J Intell Manuf 29(4):763–788

Heidelberger P (1995) Fast simulation of rare events in queueing and reliability models. ACM Trans Model Comput Simul 5(1):43–85

Hu C (1992) IC reliability simulation. IEEE J Solid-State Circuits 27(3):241–246

Iglehart DL (1975) Simulating stable stochastic systems, V: Comparison of ratio estimators. Naval Res Logist Q 22(3):553–565

Juneja S, Shahabuddin P (2006) Rare-event simulation techniques: an introduction and recent advances. Handbooks in operations research and management Sci 13:291–350

Kalchsmidt M, Verganti R, Zotteri G (2006) Forecasting demand from heterogeneous customers. Int J Oper Prod Manag 26(6):619–638

Li X, Qin J, Bernstein JB (2008) Compact modeling of MOSFET wearout mechanisms for circuit-reliability simulation. IEEE Trans Device Mater Reliab 8(1):98–121

Liang K, Jin S, Chen H, Ren J, Shen W (2020) Parametric optimization of packed bed for activated coal fly ash waste heat recovery using CFD techniques. Chin J Chem Eng 28(2):518–525

Logan DL (2012) A first course in the finite element method (5th edn), Cengage

Meketon MS, Schmeiser B (1984) Overlapping batch means: something for nothing? In: Proceedings of the 1984 WSC. IEEE

Meyn SP, Tweedie RL (2012). Markov Chains and stochastic stability. Springer

Mielczarek B, Uziałko-Mydlikowska J (2012) Application of computer simulation modeling in the health care sector: a survey. Simulation 88(2):197–216

Muñoz DF, Glynn PW (1997) A batch means methodology for estimation of a nonlinear function of a steady-state mean. Manage Sci 43(8):1121–1135

Muñoz DF, Glynn PW (2001) Multivariate standardized time series for steady-state simulation output analysis. Oper Res 49(3):413–422

Muñoz DF (2010) On the validity of the batch quantile method in Markov chains. Oper Res Lett 38(3):223–226

Muñoz DF, Muñoz DF (2011) Bayesian forecasting of spare parts using simulation. In: Altay N, Litteral LA (eds) Service parts management: demand forecasting and inventory control. Springer, pp 105–123

Muñoz DF, Muñoz DG, Ramírez-López A (2013) On the incorporation of parameter uncertainty for inventory management using simulation. Inte Trans Oper Res 20(4):493–513

Muñoz DF (2014) Input analysis for stochastic simulations. In: Wang J (ed) Encycopledia of business analytics and optimization. IGI Global, pp 1213–1223

Muñoz DF (2017) Estimation of expectations in two-level nested simulation experiments. In: Proceedings of the 29th European modeling and simulation symposium. DIME University of Genoa, pp 233–238

Nahmias S (2013) Production and operation analysis (6th edn). Irwin

Natarajan K, Pachamanova D, Sim M (2009) Constructing risk measures from uncertainty sets. Oper Res 57(5):1129–1141

Oliveira JB, Jin M, Lima RS, Kobza JE, Montevechi JAB (2019) The role of simulation and optimization methods in supply chain risk management: performance and review standpoints. Simul Model Prac Theory 92:17–44

Ostfeld A, Kogan D, Shamir U (2002) Reliability simulation of water distribution systems–single and multiquality. Urban Water 4(1):53–61

Ouyang M, Duenas-Osorio L (2012) Time-dependent resilience assessment and improvement of urban infrastructure systems. Chaos: Interdiscip J Nonlin Sci 22(3):033122

Panteli M, Mancarella P, Trakas DN, Kyriakides E, Hatziargyriou ND (2017) Metrics and quantification of operational and infrastructure resilience in power systems. IEEE Trans Power Syst 32(6):4732–4742

Pasupathy R, Schmeiser B (2010) The initial transient in steady-state point estimation: contexts, a bibliography, the MSE criterion, and the MSER statistic. In: Proceedings of the 2010 WSC, IEEE, pp 184–197

Pegden CD (2017) Deliver on your promise: How simulation-based scheduling will change your business. Simio LLC

Ramírez-Nafarrate A, Muñoz DF (2016) Performance evaluation of output analysis methods in steady-state simulations. J Comput Appl Math 301:64–73

Ren J, Gao XG, Zheng JS, Zhang Y (2010) Mission decision-making for UAV under dynamic environment. Syst Eng Electron 32(1):100–103

Ren J (2018) GM (1, N) method for the prediction of anaerobic digestion system and sensitivity analysis of influential factors. Biores Technol 247:1258–1261

Robinson S (2002) New simulation output analysis techniques: a statistical process control approach for estimating the warm-up period. In: Proceedings of the 2002 WSC. IEEE, pp 439–446

Salathe M, Bengtsson L, Bodnar TJ, Brewer DD, Brownstein JS, Buckee C et al (2012) Digital epidemiology. PLoS Comput Biol 8(7):

Schmidt JW, Taylor RE (1970) Simulation and analysis of industrial systems. Irwin

Schruben L (1983) Confidence interval estimation using standardized time series. Oper Res 31(6):1090–1108

Serfling RJ (2009) Approximation theorems of mathematical statistics. Wiley

Smith JS, Sturrock DT, Kelton WD (2019) Simio and simulation: modeling, analysis, applications (5th edn). Simio LLC

Wagner JM, Shamir U, Marks DH (1988) Water distribution reliability: simulation methods. J Water Resour Plann Manage 114(3):276–294

Wang W, Reddy V, Krishnan AT, Vattikonda R, Krishnan S, Cao Y (2007) Compact modeling and simulation of circuit reliability for 65-nm CMOS technology. IEEE Trans Device Mater Reliab 7(4):509–517

Wang RJ, Glynn PW (2016) On the marginal standard error rule and the testing of initial transient deletion methods. ACM Trans Model Comput Simul 27(1):1–30

White KP Jr (1997) An effective truncation heuristic for bias reduction in simulation output. Simulation 69:323–334

White Jr KP, Cobb MJ, Spratt SC (2000) A comparison of five steady-state truncation heuristics for simulation. In: Proceedings of the 2000 WSC, IEEE, pp 755–760

Wu D, Olson DL (2008) Supply chain risk, simulation, and vendor selection. Int J Prod Econ 114(2):646–655

Wu T, Huang S, Blackhurst J, Zhan X, Wang S (2012) Supply chain risk management: An agent-based simulation to study the impact of retail stockouts. IEEE Trans Eng Manage 60(4):676–686

Yang S, Xiao Z, Deng C, Liu Z, Zhou H, Ren J, Zhou T (2020) Techno-economic analysis of coal-to-liquid processes with different gasifier alternatives. J Clean Prod 253:120006

Chapter 7
Cause-Effect Analysis, Barriers Identification, and Policy Implications for China's Energy Security

Linmao Ma and Long Zhang

Abstract Energy is viewed as the foundation of modern industry, and energy security is particularly important for the sustainable development of the national economy. However, energy security has been affected by various factors, especially in the background of industrialization and globalization. Based on the existing research on energy security issues, the relevant factors affecting energy security are summarized in this study. In order to further investigate and explore the specific impacts of these factors on energy security, a fuzzy cognitive map has been constructed based on cognitive map theory subsequently, to quantify the impact of various factors on energy security and reveal the specific influence mechanism. Finally, by establishing five different scenarios, the effects of some key factors on the entire energy security system are analyzed and discussed. Among these factors, energy diversity, energy prices, and energy management systems are thought to have significant impacts on energy security and the development of renewable energy, and some policy implications have been provided.

Keywords Energy security · Fuzzy cognitive map · influence mechanism · Scenario simulation

7.1 Introduction

Energy plays a vital role as industrial food in economic development, especially for developing countries like China. Thus, ensuring national energy security is of great significance for social sustainability development. The concept of energy security comes with the global energy crisis or energy disputes in a specific region or within some countries, which was firstly proposed in the 1970s arising from the oil crises

L. Ma
School of Management, South-Central University for Nationalities, Wuhan 430074, China

L. Zhang (✉)
School of Business, Xinyang Normal University, No. 237 Nanhu Road, Xinyang 464000, China
e-mail: longzh@xynu.edu.cn

J. Ren (ed.), *Multi-Criteria Decision Analysis for Risk Assessment and Management*, Industrial Ecology and Environmental Management 1,
https://doi.org/10.1007/978-3-030-78152-1_7

(Blum and Legey 2012; Gasser 2020) because of the appearance of the vulnerabilities of developed economies to oil price shocks during the oil crises. In order to address this emerged issue, International Energy Agency (IEA) was constituted as the most important multinational energy platform defined this concept of energy security as the national energy system with the ability to provide reliable availability of energy supply at an affordable price (Gasser 2020; IEA. 2007; Paravantis et al. 2018). Since then, energy security has been a hot issue and gets increasing attention in various areas due to its importance to national security. What's more, based on IEA's definition of energy security, the EU and OPEC proposed a more comprehensive concept of energy security, which states that environmental issues related to energy utilization and production should be considered in defining energy security. Additional researchers give different definitions of energy security from distinctive perspectives in line with their own study. For a detailed introduction to the concept of energy security, the review on this issue conducted by Gasser can be referred (Gasser 2020). Kruyt et al. (2009) firstly investigated energy security in Europe from the perspective of availability, affordability, accessibility, and acceptability of energy systems (Kruyt et al. 2009), which is the basic of the 4-As evaluation framework (Yao and Chang 2014). Although extensive studies have shown that the evaluation model based on the 4-As framework is viewed as the core of energy security (Chester 2010; Jewell et al. 2014; Ren and Sovacool 2014), some scholars also gave their insights on this issue. Cherp and Jewell (Cherp and Jewell 2014) rightfully mentioned that the concept of energy security should deal with the following question: "Security for whom?", "Security for which values?", and "Security from what threats?" and descript it as a low vulnerability of vital energy systems. Meanwhile, Lin and Raza (Lin and Raza 2020) evaluated the level of China's energy security, which underlined the role of energy demand, energy supply, and energy security. Through a review of the contemporary literature on energy security, it has previously been observed that political and economic factors are emphasized for evaluating energy security while social and environmental factors are ignored (Novikau 2019). Furthermore, the relevant factors contain qualitative, semi-quantitative, and quantitative elements, so qualitative research dominate in the existing investigations on the relationship between these factors. What's more, with the rapid growth of economic development, China has been the largest energy consumer in the world for almost a decade, so there has been an increasing recognition that more attention needs to be paid to ensuring its energy security. However, the fact is that energy security hasn't been improved during the 30 years of reform in China (Yao and Chang 2015). Given the rising risks and shortages in energy supply, there is an essential need to enhance the study on this issue to ensure economic and social sustainability. Considering the above situation, a quantitative model is used to describe and explain the complex relationships among these various factors concerning energy security in this study.

7.2 Methods

This section aims to introduce the background of the basic information about the fuzzy set and the cognitive map theory, and then presents the process of intuitionistic fuzzy cognitive map (IFCM) in details.

7.2.1 Basic Procedure of FCM

Fuzzy cognitive map (FCM) is a popular semi-quantitative modeling approach which structure is similar to the artificial neural network, and it is usually used to describe the causal and effect relationships between factors in a complex system for its ability to deal with nonlinear problems. FCM was proposed by Kosko as a general extension of the original cognitive map (Groumpos 2010; Kosko 1986). In fact, Cognitive map (CM) is firstly presented by Axelrod in 1976 aiming at tackling complex decision-making problems in social science (Axelrod 1976), and it has been used in various fields, such as education, archaeology, planning, and urban management. CM consists of nodes and edge connections between them, and the nodes represent a set of concepts which are the variables in the real-life system, while the edges between nodes donate the casual relationships among these concepts extracting from the system. Every connecting edge with a plus or minus just only shows the positive or negative influence of one concept on another, and doesn't express the intensity of the influencing effects. Due to the nature of the original cognitive map based on modeling approach, there are some difficulties when dealing with practical problems with complicated relationships. Therefore, FCM, which inherits the basic features of CM, has been developed as an extension or improvement.

Generally, FCM is considered as a computational modeling method with the ability to simulate a complicated system involving fuzzy or uncertain situations which is depicted as a directed graph consisting of nodes and weighted arcs as shown in Fig. 7.1. These nodes donate the corresponding factors, which represent different variables, features, and states extracted from the described complex system (Amirkhani et al. 2016). Each node is assigned with a real number ranging from 0 to 1 to express the current state for a main feature in the modeled system at a specific time. The weighted edge represents the causal relationship between these factors.

Figure 7.1 is a graphic representation for a simple FCM model with 5 nodes and 8 weighted edges. In mathematical model, FCM is represented as a pair $< C, W >$ where C and W are the set of the concepts and the relationships among them, respectively. The value of concept, c_i, is within the interval of [0, 1]. The connecting edge, w_{ij}, specifies the direction and effects of the casual relationship between two concepts, c_i and c_j, and each edge is assigned with a weight value in the interval of [−1, 1], indicating the degree of the corresponding casual relationship between these concepts. Thus, W is considered as connection matrix and its formulation is expressed as Eq. (7.1) where w_{ij} represents the impact of concept c_i on concept c_j.

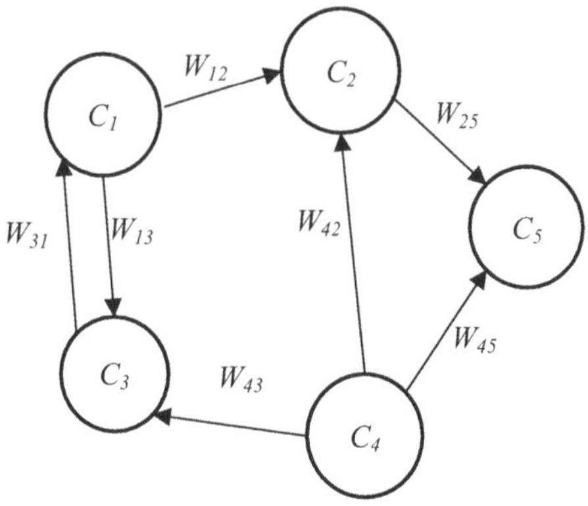

Fig. 7.1 An example structure of FCM

$$W = \begin{bmatrix} w_{11} & \cdots & w_{1j} & \cdots & w_{1n} \\ \vdots & \vdots & \vdots & \vdots & \vdots \\ w_{i1} & \cdots & w_{ij} & \cdots & w_{in} \\ \vdots & \vdots & \vdots & \vdots & \vdots \\ w_{n1} & \cdots & w_{nj} & \cdots & w_{nn} \end{bmatrix} \tag{7.1}$$

where n is the number of concepts in the practical system.

Obviously, the larger absolute value of w_{ij} represents a stronger intensity of the casual-effect relationship from concept c_i on concept c_j. The sign of w_{ij} shows different states of the causal-effect relationship between two concepts:

When $w_{ij} > 0$, the influence of c_i on c_j is positive which indicates that an increase or decrease in c_i can cause the same result in c_j;

When $w_{ij} < 0$, the causal relationship between c_i and c_j is negative which means the change of c_i will cause an opposite effect on c_j;

Whereas $w_{ij} = 0$, it suggests that there is no interconnection between c_i and c_j.

According to these rules, the connection matrix W in Fig. 7.1 can be specified as Eq. (7.2).

$$W = \begin{bmatrix} 0 & w_{12} & w_{13} & 0 & 0 \\ 0 & 0 & 0 & 0 & w_{25} \\ w_{31} & 0 & 0 & 0 & 0 \\ 0 & w_{42} & w_{43} & 0 & w_{45} \\ 0 & 0 & 0 & 0 & 0 \end{bmatrix} \tag{7.2}$$

As to FCM, during the simulations, the value of each concept at each iteration is computed based on Eq. (7.3).

$$C_i(t+1) = f\left(k_1 C_i(t) + k_2 \sum\nolimits_{\substack{j=1 \\ j \neq i}}^{n} C_j(t) w_{ji}\right) \tag{7.3}$$

where t is the simulation iteration, n is the amount of concept and $f(x)$ is the threshold function. Accordingly, $C_i(t+1)$ and $C_i(t)$ are the value of concept c_i at the $(t+1)$th and tth iteration, respectively, and then C_j is the value of concept c_j at the tth iteration. Generally, the value of $k_1 = k_2 = 1$ is assumed.

The simulation procedure will continue until the condition $C(t+1) = C(t)$ or $C(t+1) - C(t) \leq e$ is met, where e is the residual determined by specific problems. It indicates that the modeled system has gotten to a steady-state when the condition is satisfied, and the final values of all concepts can be achieved.

There are several types of threshold function $f(x)$ can be used (Bueno and Salmeron 2009):

$$\text{Bivalent } f(x) = \begin{cases} 0 \ x \leq 0 \\ 1 \ x > 0 \end{cases} \quad (7.4)$$

$$\text{Trivalent } f(x) = \begin{cases} -1 \ x \leq -0.5 \\ 0 \ -0.5 < x < 0.5 \\ 1 \ x \geq 0.5 \end{cases} \quad (7.5)$$

$$\text{Sigmoid } f(x) = \frac{1}{1 + e^{-\alpha x}} \quad (7.6)$$

In Eq. (7.6), $\alpha > 0$, is a vital parameter for this function which determines the steepness of the threshold function. In most cases, the sigmoid function is used as the threshold function so as to restrict the value of all factors in the universe [0,1] (Amirkhani et al. 2016; Han and Deng 2018) and the value of α is usually set to 1, which making this function approximate to a linear one and outperformed the other functions in the literature (Bueno and Salmeron 2009; Papageorgiou et al. 2011).

According to the description mentioned above, we can outline the basic procedure of FCM (Groumpos 2010):

Step 1: Inviting experts to identify the factors relevant to the modeled system.

In this stage, a group of experts who have great understandings on this issue are invited to determine which element comprise the FCM model based on their own experience or knowledge, and then all experts may discuss about these elements to identify the final elements to construct the FCM model which can be accepted by all experts. Furthermore, the interrelation among these elements is described by the experts. They point out whether there is a causal and effect relation between two elements or not and the direction of the causality if existing. Hence, the basic structure of FCM model is constructed.

Step 2: Determining initial value to the concept.

For the constructed FCM model, the initial values of all concepts represent the present state of this system at the given time and can be assigned in different approaches:

extracting from historical data or given by experts. As to the former way, the fuzzification should be applied to these values. If the latter one is adopted, the experts should describe the current state of all concepts in linguistic variables. Then, these linguistic values are defuzzied into a crisp value.

Step 3: Defining the intensity of the interrelation.

The weight values represent the intensity of the relation among these concepts, and are given by the experts. Similar to the description in Step 2, the experts give their suggestions on the degree of the relation in linguistic variables, and then these linguistic variables are transformed into crisp numbers by using the defuzzification method of center gravity.

Step 4: Simulating the inference processes.

During the simulations, the values of concepts are calculated by using Eq. (7.3), and then the new values repeatedly replace the values in the last iteration.

Step 5: Checking the termination condition.

If the condition $C(t+1) = C(t)$ or $C(t+1) - C(t) \leq e$ is satisfied, the final values of concepts are outputted for the further decision-making. Otherwise, it goes back to continue Step 4 until the above conditions are satisfied.

7.2.2 *Intuitionistic Fuzzy Set Theory*

In practice, human judgments cannot describe their opinions on a given issue exactly and detailly in most cases, due to a lack of complete or abundant information even though they are experts with expertise knowledge associated with this issue. Therefore, there exist vagueness, uncertainty, and hesitation in their decisions during the decision-making process, especially when involving multiple objectives of complicated systems. Uncertain information often appears in multi-criteria decision-making, and the uncertainty of information will seriously affect the accuracy of decision-making. There is an urgent need to deal with fuzzy and uncertain information in a complex system, especially for large-scale problems. So it's in this context that the fuzzy set (FS) theory was developed to deal with fuzziness and the partial truth of information (Zadeh 1965) which is widely used in various fields for solving practical problems with fuzzy constraints and then has been well developed by a large number of scholars. Subsequently, intuitionistic fuzzy set (IFS) theory was proposed as an expanded fuzzy theory (Atanassov 1986, 1999), besides the definition of membership function given in traditional fuzzy logic, the non-membership function of each element in this set is also given. Let's assume that Set X is a nonempty fixed set of the universe, and then an IFS A in X is expressed as:

$$A = \{\langle x, \mu_A(x), \nu_A(x)\rangle | x \in X\} \quad (7.7)$$

where $\mu_A(x) : X \to [0, 1]$ is the membership function of element $x \in X$ to the set $A \subseteq X$ while $v_A(x) : X \to [0, 1]$ is the corresponding non-membership function. $\mu_A(x)$ and $v_A(x)$ meet the condition: $0 \le \mu_A(x) + v_A(x) \le 1$. According to the formulation of intuitionistic fuzzy sets, we can see that traditional fuzzy set is actually a special presentation of traditional fuzzy set. Especially, the IFS is reduced to the output of a traditional FS when $\mu_A(x) + v_A(x) = 1$. In addition to the membership and non-membership function, for an IFS, there is another concept of hesitancy degree $\pi_A(x) = 1 - \mu_A(x) - v_A(x)$ where $0 < \pi_A(x) < 1$ which represents the uncertainty or hesitancy of whether x belongs to the set A. Thus, IFS has the ability to handling much of the hesitancy in complex circumstances compared with traditional FS. Therefore, the trapezoidal intuitionistic fuzzy number A is expressed $A = \langle (a_1, a_2, a_3, a_4), (b_1, b_2, b_3, b_4) \rangle$, and the membership and non-membership functions are shown as Eq. (7.8) and (7.9). The trapezoidal fuzzy number is transferred into a triangle fuzzy number under a special situation where the most promising value is equal, namely $a_2 = a_3$ and $b_2 = b_3$.

$$\mu_A(x) = \begin{cases} 0 & x < a_1 \\ \frac{x-a_1}{a_2-a_1} & a_1 \le x \le a_2 \\ 1 & a_2 \le x \le a_3 \\ \frac{a_4-x}{a_4-a_3} & a_3 \le x \le a_4 \\ 0 & x > a_4 \end{cases} \tag{7.8}$$

$$v_A(x) = \begin{cases} 1 & x < b_1 \\ \frac{x-b_1}{b_2-b_1} & b_1 \le x \le b_2 \\ 0 & b_2 \le x \le b_3 \\ \frac{b_4-x}{b_4-b_3} & b_3 \le x \le b_4 \\ 1 & x > b_4 \end{cases} \tag{7.9}$$

7.2.3 *Intuitionistic FCM*

According to the description above, FCM is a suitable technique for modeling complex systems involving various elements (concepts, variables, factors, or characteristics), because it can depict and quantify the interrelation among these concepts rather than just conduct a simple pairwise comparison. What's more, substituting the traditional FS with IFS is a feasible approach to improve the experts' ability to tackle the vagueness of information in complex practical circumstances. Therefore, IFCM, which integrates IFS with FCM, is an effective way for all stakeholders to evaluating the current states of the concepts and the corresponding interrelations among them (Dursun and Gumus 2020). The procedure of the intuitionistic fuzzy cognitive map is shown in Fig. 7.2, and the detailed steps of this IFCM-based approach are as follows (Dursun and Gumus 2020):

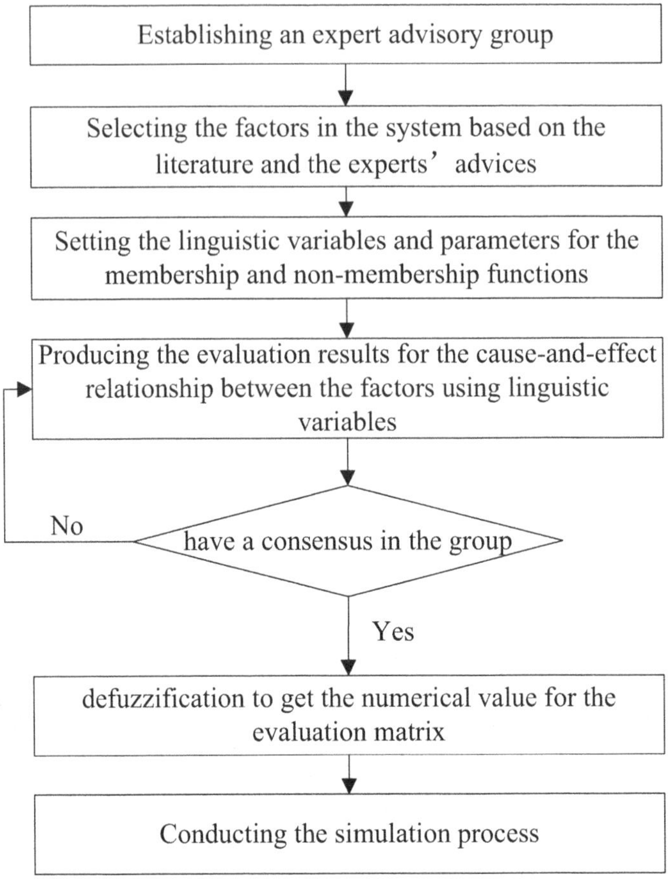

Fig. 7.2 The process of intuitionistic fuzzy cognitive map

Step 1: Inviting all experts from various fields related to energy security, including official staffs from government administration, experts from enterprises and institutes, as well as individuals randomly selecting from common consumers, to form a group of advisory committee. Based on the literature review and the committee's opinions on the factors related to the given issue, we can determine the concepts that construct the FCM. By doing this, it provides a convenient way that integrates both the previous knowledge and individual perceptions, and improves the participation of all participants' cognition of this specific issue.

Step 2: Defining the linguistic variables and the parameters of the membership and non-membership functions. There are two categories of linguistic variables used for evaluating the current state of the concepts and the degree of the causality among these concepts, respectively.

Table 7.1 Linguistic variables of fuzzy numbers for initial state

Linguistic variables	Score	Intuitionistic trapezoidal fuzzy numbers
Very good	4	⟨(1, 1, 1, 1), (1, 1, 1, 1)⟩
Good	3	⟨(0.7, 0.8, 0.9, 1), (0.7, 0.8, 0.9, 1)⟩
Medium	2	⟨(0.3, 0.4, 0.5, 0.6), (0.2, 0.4, 0.5, 0.7)⟩
Poor	1	⟨(0, 0.1, 0.2, 0.3), (0, 0.1, 0.2, 0.3)⟩
Very poor	0	⟨(0, 0, 0, 0), (0, 0, 0, 0)⟩

The first type of linguistic variables is a five-scale one and used to evaluate the initial states of all the concepts. Table 7.1 shows the linguistic variables and the corresponding intuitionistic trapezoidal fuzzy numbers in details. The experts give their opinions on the current states of each element in the modeled system. Similarly, the experts' determination on the degree of the influence of one element on another is acquired by using another type of seven-scale linguistic variable, and the corresponding fuzzy numbers are as described in Table 7.2. The membership and non-membership functions with respect to these two types of fuzzy numbers are depicted in Fig. 7.3. It is worth noting that the experts have reached a consensus on the direction of all influences in the system before they determine the corresponding degree.

Step 3: Computing the crisp values of the initial states and the connect matrix (i.e., weight matrix). After the experts' opinions on the initial states and the intensity of the influence are submitted, all the fuzzy initial states are aggregated to produce an overall fuzzy number for all the concepts. In the same way, the overall fuzzy connect matrix is generated, and we have assumed the same weight for all experts' suggestions in calculating the overall fuzzy numbers for the initial states and the intensity of the influencing effects. After that, these fuzzy numbers are transformed into crisp values by means of the defuzzification method of the centroid of gravity. Eventually, the numerical state, belonging to the interval [0,1], and connection matrix, belonging to [−1,1], are yielded, respectively.

Table 7.2 Linguistic variables and fuzzy numbers for the degree of influence

Linguistic variables	Score	Intuitionistic trapezoidal fuzzy numbers
Absolutely strong influence	6	⟨(1, 1, 1, 1), (1, 1, 1, 1)⟩
Strong influence	5	⟨(0.7, 0.8, 0.9, 1), (0.7, 0.8, 0.9, 1)⟩
Fairly strong influence	4	⟨(0.5, 0.6, 0.7, 0.8), (0.4, 0.6, 0.7, 0.9)⟩
Medium influence	3	⟨(0.3, 0.4, 0.5, 0.6), (0.2, 0.4, 0.5, 0.7)⟩
Fairly weak influence	2	⟨(0.1, 0.2, 0.3, 0.4), (0.1, 0.2, 0.3, 0.5)⟩
Weak influence	1	⟨(0, 0.1, 0.2, 0.3), (0, 0.1, 0.2, 0.3)⟩
Absolutely weak influence	0	⟨(0, 0, 0, 0), (0, 0, 0, 0)⟩

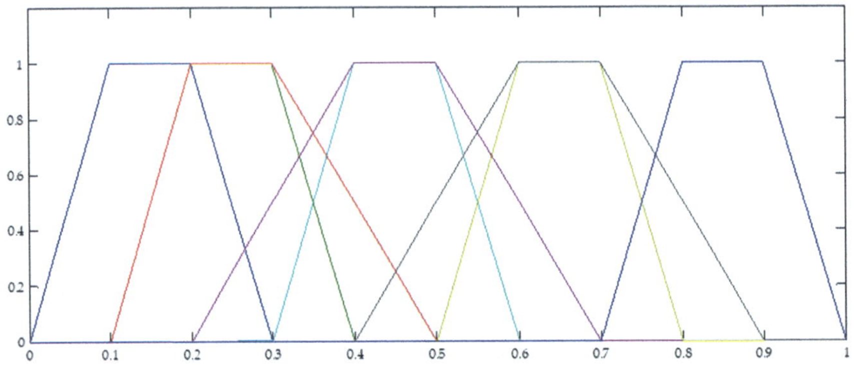

Fig. 7.3 Membership and non-membership functions of trapezoidal IFN

Step 4: Executing the simulation process. In this stage, the simulation refers to the reasoning process of FCM and can be executed according to the aforementioned calculation rules based on the gained initial value of the concept and connect matrix. Finally, intuitionistic FCM yields the last states of all concepts.

Step 5: Conducting the scenario simulations. As to the constructed intuitionistic FCM, the last state of the modeled system can be analyzed under different conditions with scenario simulations, because the structure of FCM allows interaction with external variables at each iteration. We can get some useful information through the comparison between the last states that the system yields under different conditions with that in the business-as-usual scenario.

7.3 Factors Identification

According to the current investigations on energy security, the corresponding influential factors are re-identified and summarized in this paper in line with energy supply and consumption including commercial and civil users. Finally, we have identified 15 factors that may affect energy security, which is as follows:

Self-sufficiency.

China has accounted for 24% of global primary energy consumption and 34% of energy consumption increment in 2018 (British Petroleum 2020). In fact, this country has already become a major player in international energy market and leaded global crude oil imports since 2017. Presently, China is highly dependent on oil import, which has increased by 8.4% in 2018, and the oil import dependence has even reached a historical record of 72% of its crude oil consumption. Meanwhile, China's natural gas imports have also surpassed Japan in 2018, becoming the largest natural gas importer around the world, and the external dependence on natural gas was as high as 45.3% (CPEA 2019). The high dependence on various imported primary energy has greatly affected its self-sufficiency of energy supply, and posed a serious threat to energy supply security and stability (Zha 2006).

Energy diversity.

Energy diversity refers to the multiple and diverse energy sources in the national or regional energy mix. Usually, energy diversity is affected by factors such as resource endowment, geographic location, mining technology level, transportation, and energy policy. Diversified energy sources are helpful in improving energy resilience, which can optimize the structure of the energy market and reduce the risks arising from the overdependence on energy imports. Therefore, achieving energy diversification is beneficial to the enhancement of energy security (Vivoda 2009).

Renewable energy production.

The development of renewable energy is an effective way to promote the sustainable development of the national energy system, and it also determines the safety of the energy supply (Hache 2018). Renewable energy sources, such as hydro, wind, solar and biomass, as promising alternatives to traditional energy, can alleviate the environmental pollution problems caused by the consumption of traditional fossil fuels, have great potentials in making up for the shortage of oil and natural gas resources, increasing energy supply, meeting energy demand, and reducing greenhouse gas emissions. Therefore, promoting renewable energy development is an important way to enhance energy security. Due to the important role of renewable energy in enhancing regional energy security, it is necessary to accelerate the investment and development of renewable energy resources (Bekhrad et al. 2019).

Industrial structure.

Factors related to the industrial structure have a great influence on the demand for energy resources, and a few heavy industrial sectors have contributed a quite share in total energy consumption (Chen et al. 2014). Therefore, the adjustment of the industrial structure can have a great impact on China's energy system. The adjustment and optimization of industrial structure can reduce the dependence of the energy system on traditional fossil fuels to alleviate a series of social problems caused by the shortage of fossil fuels. Reasonably adjusting the industrial structure, appropriately increasing the proportion of the tertiary industry, and reducing the share of the secondary industry in the development of the national economy can effectively alleviate the shortage of energy supply and promote the green supply of energy and low-carbon development.

Technological improvement and maturity.

Technology has been an increasingly important factor in energy security (Sovacool and Mukherjee 2011), and the improvement and maturity of technology can affect both energy supply and demand, which is reflected in two aspects. On the one hand, the improvement of traditional technologies and equipment can improve the efficiency of energy utilization to reduce energy intensity in economic production. On the other hand, the innovation of clean and renewable energy technologies can provide alternative energy sources to enhance energy diversity and reduce environmental pollutions. The innovation in the energy areas involves the advancement in technology related to energy production, transportation, conversion, and utilization.

Price stability of energy products.

As energy products and services is a major commodity in international trade, a stable energy price (i.e., the price of crude oil, natural gas, coal, gasoline, and electricity) is very important to ensure energy supply security and can affect the stability of the relationship between energy supply and demand in the energy market (Sovacool and Mukherjee 2011). Compared with general product prices, energy prices are affected by multiple factors such as geopolitics, international situations, natural environment, national policies, and transportation channels (Criekemans 2011). Especially, for those countries or areas that are not capable of participating in shaping the global energy pricing system, it would be very passive and make it difficult to achieve cost optimization for importing energy. Finally, energies imported at excessively high prices may even affect economic security and sustainability.

Decentralization and Electrification.

This element measures the ability of end-users accessing to reliable and available grid connections. The increase of decentralization and electrification can improve energy efficiency and reduce the consumption of other energy sources (Ren and Sovacool 2014). What's more, the decentral and smart grid can improve the resilience of the power system, help more people gain access to the energy system and enhance national energy security, which is of greater significance for the residents in the rural and remote areas.

Environmental sustainability.

For a long time, China, likes other developing countries, has paid more attention to economic growth rather than environmental issues, and the initial concept of energy security defined by IEA also doesn't contain any environmental sustainability elements. However, environmental factors indeed have a significant impact on the development of the energy system because of the growing concerns of salient climate change and increasing air pollution issues from both national and international aspects (Sovacool 2014). Additionally, energy consumption, mostly fossil fuels, has contributed greatly to both economic growth and pollutant emissions in most developing countries, so how to achieve the balance between economic growth and environmental protection is an actual problem in front of all countries (Zeng et al. 2011).

Social perception.

The social perception denotes the public awareness and attitudes toward the current energy system and the government's energy plans (Wüstenhagen et al. 2007). Public understanding and awareness of the energy system may affect their energy consumption behaviors as well as their participation and support in the implementation of the government's energy policy.

Transparency of energy policy.

The transparency of energy policy indicates the publicity of the national energy development plan and various energy policies, and determines the public's awareness of the national energy system and future development goals (Mendonça et al. 2009). The more thorough the public's understanding of various energy policies, the more they will cooperate with the national energy development plan, to achieve the promotion of energy system reform and enhance the resilience of the energy system.

Energy efficiency.

Energy efficiency measures the ratio between the total economic output and energy input in energy production, conversion, and utilization of various energy sources (Zhou and Ang 2008). With the improvement of energy efficiency, less energy will be consumed, and more economic output will be generated, so that energy security can be enhanced.

Import stability.

Energy import stability refers to the way and ability to obtain necessary energy products and services from other countries or regions, which is very important for countries with a high external dependence on energy imports. Once the supply of imported energy is regularly disturbed by external forces, it may pose great pressures and even threats to national energy security. As a major role in the international energy market, China is increasingly depending on imported energy sources to meet domestic energy supply, especially that its oil and natural gas imports have already reached the historically high level, and are very likely to keep this high level in the following decades, which makes it vulnerable to the instability of energy import (Gnansounou 2008).

Political stability.

A stable political environment is viewed as a prerequisite to ensure the long-term development of the whole energy system (Jacobsson and Lauber 2006). Political stability, especially the consistency and continuity of energy policies, including various energy plans constructed by the central and local governments and the corresponding safeguard measures, may have a fundamental effect on the investment in the energy industry, especially for renewable energy companies. The stable and coherent policies can enhance investors' confidence in the energy projects and facilitate the implementation of the government's corresponding energy policies.

Military power.

Military power in this study mainly refers to the ability to ensure the safety of overseas energy imports. For countries with a higher energy external dependence, national military power is a necessary condition to ensure the safety of energy imports, and it can help to deal with the potential risks in international energy trading and transportation effectively (Klare 2006).

Safety and reliability.

Safety and reliability are mentioned to indicate the ability of the national energy system to withstand potential damage risks (including natural disasters and human factors), and the ability to recover after damage, which is very important for countries with rapid economic development. A stable energy system can provide a suitable environment for economic development. Meanwhile, a reliable energy system can also reduce energy consumption during its maintenance.

7.4 Scenario Simulations

According to the above-mentioned influencing factors, this section constructs the corresponding FCM. In fact, factors affecting energy security may vary from one country to another. In order to analyze these factors affecting energy security, many scholars have conducted studies on this issue, yet most of them are simple descriptive statistical analyses for the factors. In order to explain the interrelationship between these factors and show the influence mechanism, an IFCM is used to simulate the causal relationship between these factors, reveal the influence mechanism, and provide corresponding solutions from different perspectives.

7.4.1 Construction of IFCM

Based on the summary of the existing literature, scholars in related fields are invited to form an expert advisory group to reconfirm each factor. The concepts used to construct the IFCM are shown in Table 7.3. Then the expert group evaluates the causal relationship between the factors, and finally it generates the original FCM as shown in Fig. 7.4, where blue edges represent positive causality, and red edges represent negative causality.

Based on the initial fuzzy cognitive map in Fig. 7.4, all members in the expert group use the language variables to describe the intensity of the causal relationships among the factors, as shown in Table 7.4. When all experts give their advice on the relationships, the de-fuzzy operation is implemented to yield the final weight matrix, as shown in Table 7.5. Finally, the intuition fuzzy cognitive map is implemented on the MATLAB platform.

In this section, the IFCM established by the expert group is used to analyze the cause-and-effect relationships among the various factors affecting national energy security from the perspective of improving energy security and investigating the impact of different factors on the entire system in different scenarios. Furthermore, China's energy security is explained in detail under different energy policies and development paths.

Table 7.3 Summary of factors affecting energy security

Factor	Content
F1: Energy security state	It indicates the assessed energy security level and the possible states of energy security level under different scenarios
F2: Self-sufficiency	It measures the ability of a country or region to meet its energy demand with domestic energy production. It corresponds to the degree of external dependence on energy, and the higher self-supply level means lower dependence on external energy
F3: Energy diversity	It shows the ability to meet energy demand with different energy sources, and multi-channel energy supply can help to enhance security of energy systems
F4: Renewable energy production	The aggressive development of renewable energy is helpful to alleviate the environmental problems caused by the massive consumption of traditional fossil fuels and provide alternative energy supply to make up for the country's restrictions on traditional fossil energy
F5: Industrial structure	It represents the rationality of national or regional industrial structure as well as its potential impact on energy security by considering the energy demand in different industrial sectors
F6: Technological improvement and maturity	It reflects the development of existing energy-related technologies which determines the efficiency and possibility in energy production, conversion, and consumption
F7: Price stability	This factor has an obvious effect on energy trading. Different from prices of general products, energy prices can be affected by various factors such as politics, natural environment, etc. High-priced imported energy products may cost a lot of foreign exchange and even affect energy importers' economic and energy security
F8: Decentralization and Electrification	It denotes the application of distributed energy systems in the entire energy system and the degree of residents' access to a reliable grid network
F9: Environmental sustainability	It reflects the impact of energy production and utilization on the environment, and the energy-related environmental challenges can be more series in large developing countries and may have a fundamental influence on the government's energy plans

(continued)

Table 7.3 (continued)

Factor	Content
F10: Social perception	This factor refers to public awareness of national or regional energy system as well as government's planning for energy development
F11: Transparency of energy policy	It measures public's understanding of various energy policies and energy development plans
F12: Energy Efficiency	It denotes the level of transforming energy inputs into economic output with current energy technologies. The improvement of energy efficiency is beneficial to save energy and reduce pollution emissions
F13: Import stability	It reflects the ability to obtain energy sources from the foreign countries
F14: Political stability	It measures the inherent state of national and local energy-related policies
F15: Military power	It illustrates the ability to enhance the safety of purchasing and transporting energy resources from overseas
F16: Safety and reliability	It demonstrates the resilience or ability against various risks of the national energy system

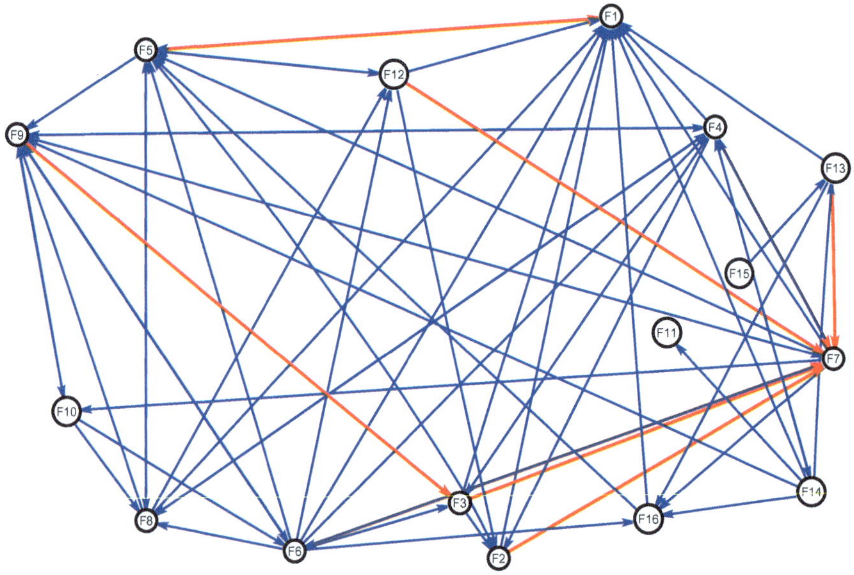

Fig. 7.4 The initial FCM for energy security

Table 7.4 The causal relationship between the factors in language variables

	F1	F2	F3	F4	F5	F6	F7	F8	F9	F10	F11	F12	F13	F14	F15	F16
F1	VL	VL	L	VL	VL	VL	VL	VL	VL	VL	VL	VL	VL	VL	VL	VL
F2	FH	VL	−L	−L	VL	VL	L	VL	VL	L	VL	L	VL	VL	VL	VL
F3	VL	VL	VL	VL	L	VL	L	M	M	VL	VL	M	−FL	−FL	−FL	−M
F4	VL	VL	VL	VL	L	L	L	FH	H	FL	M	L	−L	−L	−L	−L
F5	VL	L	L	VL	VL	L	L	VL	VL	VL	L	VL	VL	VL	VL	VL
F6	VL	L	VL	VL	VL	VL	VL	VL	VL	VL	VL	VL	VL	VL	VL	VL
F7	VL	VL	L	VL	L	L	VL	L	L	L	L	VL	VL	VL	VL	L
F8	M	VL	L	L	VL	L	VL	VL	L	VL	VL	L	−L	VL	−L	VL
F9	FH	VL	L	L	−L	FL	VL	−FL	VL	FL	VL	L	−L	−L	−FL	−L
F10	L	VL	VL	VL	VL	VL	FL	FH	M	VL	VL	VL	VL	VL	VL	VL
F11	L	L	L	−L	VL	VL	VL	VL	VL	L	VL	VL	VL	VL	VL	VL
F12	L	M	VL	VL	L	L	VL	M	FH	VL	VL	VL	VL	VL	VL	VL
F13	−VH	−FH	−FL	−FL	−L	−M	−M	−M	−H	−M	VL	−FH	VL	VL	VL	VL
F14	−H	−FL	−L	−L	VL	−L	−L	−L	VL	−L	VL	−L	VL	VL	VL	VL
F15	−FH	−L	−L	−L	VL	−L	−L	−L	VL	−L	VL	−L	VL	VL	VL	VL
F16	−H	−M	L	L	VL	VL	VL	FL	L	L	−M	VL	VL	VL	VL	VL

Table 7.5 The final weight matrix

	F1	F2	F3	F4	F5	F6	F7	F8	F9	F10	F11	F12	F13	F14	F15	F16
F1	0.00	0.00	0.00	0.00	0.13	0.00	0.00	0.00	0.00	0.00	0.00	0.00	0.00	0.00	0.00	0.00
F2	0.30	0.00	0.00	0.00	0.10	0.00	−0.18	0.00	0.00	0.00	0.00	0.00	0.00	0.00	0.00	0.00
F3	0.34	0.10	0.00	0.00	0.00	0.00	−0.15	0.00	0.00	0.00	0.00	0.00	0.00	0.00	0.00	0.00
F4	0.24	0.23	0.45	0.11	0.00	0.00	−0.10	0.20	0.10	0.00	0.00	0.00	0.00	0.10	0.00	0.00
F5	−0.23	0.00	0.00	0.00	0.00	0.00	0.00	0.12	0.03	0.00	0.00	0.11	0.00	0.00	0.00	0.00
F6	0.01	0.00	0.12	0.10	0.20	0.00	−0.05	0.03	0.13	0.00	0.00	0.25	0.00	0.00	0.00	0.21
F7	0.20	0.00	0.00	0.11	0.25	0.13	0.00	0.00	0.00	0.31	0.00	0.00	0.00	0.00	0.00	0.11
F8	0.11	0.00	0.00	0.20	0.19	0.00	0.00	0.00	0.15	0.00	0.00	0.12	0.00	0.00	0.00	0.00
F9	0.00	0.00	−0.10	0.13	0.00	0.07	0.22	0.00	0.00	0.24	0.00	0.00	0.00	0.00	0.00	0.00
F10	0.00	0.00	0.00	0.00	0.00	0.12	0.00	0.20	0.20	0.00	0.00	0.00	0.00	0.00	0.00	0.00
F11	0.00	0.00	0.00	0.00	0.00	0.00	0.00	0.00	0.00	0.00	0.00	0.00	0.00	0.00	0.00	0.00
F12	0.12	0.21	0.00	0.00	0.12	0.00	−0.05	0.00	0.00	0.00	0.00	0.00	0.00	0.00	0.00	0.00
F13	0.32	0.00	0.00	0.00	0.00	0.00	−0.02	0.00	0.00	0.00	0.00	0.00	0.00	0.00	0.00	0.20
F14	0.01	0.00	0.00	0.38	0.00	0.00	0.00	0.00	0.03	0.00	0.45	0.00	0.37	0.00	0.00	0.11
F15	0.00	0.00	0.00	0.00	0.00	0.00	0.00	0.00	0.00	0.00	0.00	0.00	0.15	0.00	0.00	0.00
F16	0.08	0.00	0.00	0.00	0.20	0.00	0.00	0.00	0.00	0.00	0.00	0.00	0.00	0.00	0.00	0.00

7.4.2 Scenario Descriptions

Firstly, in the business-as-usual scenario, the initial state of all factors is set as 0.1, and the entire system runs without any additional constraints until getting into a basic steady-state. Secondly, in order to investigate the impact of various development paths on energy security, and to specifically study the impact of corresponding factors on other factors in the system as well as the entire system, we constructed five scenarios for analysis and comparisons. The five comparative scenarios proposed in this study correspond to different energy system development paths with specific strategies, which also focus on the relevant factors related to the distinct circumstances. By comparing the steady-state of the system in different scenarios with the result of the business-as-usual scenario, the impact of different development paths on energy security is illustrated.

The five scenarios proposed in this study are: green development, import security, energy supply reform, energy consumption reform, and public participation, and some key or focused factors are identified for each scenario, as shown in Table 7.6. In each scenario, the initial states of the focused factors are changing while the initial values of other factors are consistent with that in the business-as-usual scenario. The optimal case is that the initial states of the focused factors take values of 1, and the worst case is that the corresponding factors are set as 0.1.

S1: Green development

As to the green development scenario, the focus of the government's energy development plan is to promote the development of renewable energy and the development and use of related technologies, while emphasizing the negative impact of energy consumption on the environment. Therefore, the key factors for implementing the energy system development path with the theme of green development include:

Table 7.6 Five scenarios and the key factors in each scenario

Scenario	Key factors
S1: Green development	F4: Renewable energy production F6: Technological improvement and maturity F9: Environmental sustainability
S2: Import security	F13: Import stability F15: Military power
S3: Energy supply reform	F2: Self-sufficiency F3: Energy diversity
S4: Energy consumption reform	F5: Industrial structure F8: Decentralization and electrification F12: Energy efficiency
S5: Public participation	F10: Social perception F11: Transparency of energy policy F14: Political stability

renewable energy production (F4), technological improvement and maturity (F6), and environmental sustainability (F9).

Under the green development scenario, the initial status values of the various focused factors regarding renewable energy development are 1, which indicates that a relatively complete policy system, including related energy plans, laws, and regulations, has been established and various financial and non-financial supports have been provided to promote renewable energy development. Moreover, a large amount of publicity is issued to the residents to enhance residents' understanding of renewable energy, including the full explanation of the technical factors, potential risks, and essential measures for dealing with the difficulties in the implementation of renewable energy projects. Meanwhile, it has also strengthened the restrictions on various pollutant emissions caused by energy consumption. The worst case for this scenario shows that there is a lack of the government's investment in the promotion of renewable energy development, especially in the research and development of renewable energy-related technologies, and the government failed to take any measures to deal with the environmental issues.

S2: Import security

For the import security scenario, it mainly focuses on the security issues related to importing energy from other countries. Due to the seasonality and volatility of renewable energy sources, it is not possible for renewables to completely replace traditional fossil fuels in the short term. In order to ensure the normal development of the national and regional economy, it is particularly important to ensure the safe supply of energy imports for countries with relatively high dependence on foreign oil and natural gas. Under this scenario, import stability (F13) and military power (F15) will be the key factors in the energy development strategy.

In this scenario, the optimal situation is that the initial states of the focused import security factors are set as 1, indicating that there are sufficient external energy sources in the international energy market, and they can be conveyed to domestic with through reliable and affordable transportation modes and routes, as well as essential military powers to protect the procurement and transport of energy products when facing potential attacks or risks. On the contrary, in the worst case, it represents energy import is facing serious difficulties and obstacles due to geopolitical and international restrictions.

S3: Energy supply reform

Given the steady-self of the energy system, domestic energy supply and demand factors must be considered. Therefore, as for the energy supply scenario, energy supply is taken as a vital part of the energy strategy. Since the capacity of energy supply directly affects the stability of the national energy system, so only through the establishment of a reliable energy supply system can it ensure the normal operations of social functions. Thus, the vital factors in this scenario are self-sufficiency (F2) and energy diversity (F3).

Similarly, for the energy supply-side reform scenario, the optimal situation means that complete energy independence and diversified energy mix have been achieved, and the domestic energy supply from various sources, can satisfy its energy demand without import relying on imported energy sources. On the contrary, in the worst case, the values of self-sufficiency (F2) and energy diversity (F3) are set as 0.1 in the simulation, which implies that the domestic energy supply is in great shortage and highly relied on certain energy sources.

S4: Energy consumption reform

Energy demand is another essential component closely related to energy supply, energy demand-reform focuses on the structure and efficiency of energy conversion and consumption. Concerning the energy demand-side reform scenario, energy strategies should stress the following factors: industrial structure (F5), decentralization and electrification (F8), and energy efficiency (F12). Consequently, energy development plans bring about the promotion and improvement of energy utilization.

In this scenario, the most ideal situation for the energy system is that the industrial structure has been upgraded and optimized, so the structure of energy demand has become more reasonable, and then the energy-related technologies have been advanced to improve energy efficiency to a really high level. Moreover, the distributed energy system is widely used and electrification has been improved in the current energy system, so there are more flexible ways to access the grid network to ensure people in remote areas get more access to modern and affordable energy products and services. Accordingly, we assume that the corresponding factors have a value of 1 in the system. On the contrary, in the worst case, the industrial structure is highly energy-depended, the energy efficiency is relatively low with less economic outputs than expected, and the domestic energy network failed to be capable of converting and utilizing energy effectively, especially for remote areas where there are few accesses to modern energy products.

S5: Public participation

Finally, as far as the public participation scenario is concerned, the crucial aspect is toward the public awareness of the energy system, especially for the energy policy. Generally speaking, residents' views on energy systems are regarded as a trivial aspect compared with other factors for a centralized system of energy management, yet the improvement of public awareness to energy development is going to be stressed in the future for better implementation of energy development plans, because public or citizens have played an increasingly important role in the development of national energy system. Therefore, social perception (F10), transparency of energy policy (F11) and political stability (F14) are emphasized in this scenario.

Regarding this scenario, the parameters are set in a way similar to the above scenarios. In order to fulfill our objective of investigating how public awareness influences the energy system, factors affecting common residents' understanding and acceptance of the national energy network construction and policies are mainly emphasized. Hence, the best case denotes that the residents are supposed to have a detailed understanding of energy network construction, support national energy development plans, and be familiar with relevant energy policies and regulations so that they can cooperate with the local officials in the implementation of the national energy plans. Oppositely, in the worst case, the values of related parameters are assumed to be 0.1, which indicates that residents do not have much knowledge on the national energy network and its development plans as well as policies.

In addition to the best and worst cases, there are some values for the intermediate states separately. By comparing the stable states of the system under different conditions with the original baseline scenario, the cause-effect relationship between these factors in the system can be analyzed and discussed.

7.4.3 Simulation Results

Based on the above scenarios, the intuitionistic fuzzy cognitive map for energy security is put into running with different parameters in the scenarios until getting into a stable state. Through the simulation results, the impact of the specific factors concerned in various situations on the system is discussed as follows.

Regarding the green development scenario, the results are shown in Fig. 7.5, and it illustrates the effects of the elements related to renewable energy development. The results reveal that these focused elements in this scenario have a significant influence on the energy security system, and the other factors fluctuate with the change of renewable energy development. Notably, under the worst condition, energy diversity (F3) of the energy system has experienced a decrease of more than 10% compared to the steady-state of business-as-usual scenario, and self-sufficiency (F2), decentralization and electrification (F8), and energy efficiency (F12) have a reduction of more than 5%. On the contrary, for the optimal case, energy diversity (F3) and energy efficiency (F12) increase by over 8%, and self-sufficiency (F2) and social perception (F10) also rise approximately 5%, respectively. On the whole, energy security statue (F1) and safety and reliability (F16) have an increment of nearly 5%. The compared results demonstrate that renewable energy development can update the

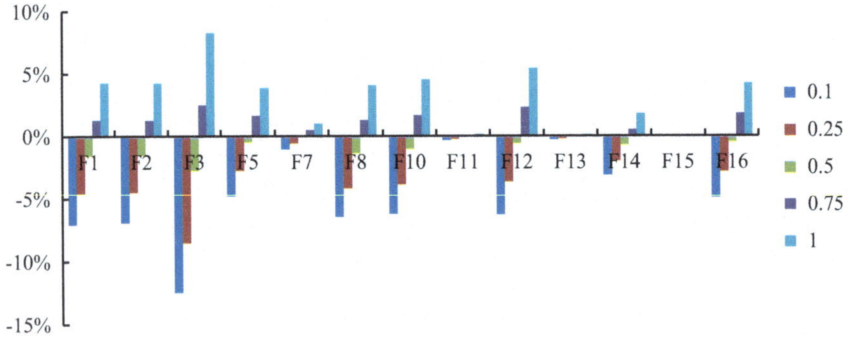

Fig. 7.5 System variations in the green development scenario

national energy system through the application of novel technologies and the advance of public awareness.

Similar to the green development scenario, the results of the import security scenario stressing the effect of factors regarding energy importation are pictured in Fig. 7.6. Obviously, import stability (F13) and military power (F15) have significant impacts on energy security statue (F1) and safety and reliability (F16) of the energy system, while other elements in the system barely fluctuate with these two factors. Hence, the energy strategy just taking energy import into account is relatively independent from other development paths.

Furthermore, as to the energy demand-side reform scenario, the change of the factors in the system is shown in Fig. 7.7. Three factors, self-sufficiency (F2), renewable energy production (F4), and decentralization and electrification (F8), have decreased by more than 4% in the worst situation and conversely increased by varying percent under the optimal circumstance, which indicates that the adjustment and reform of energy consumption have a positive effect on the energy system, and promote renewable energy development and electrification in the rural area. What is more, the more than 4% of growth in energy security (F1) for the optimal

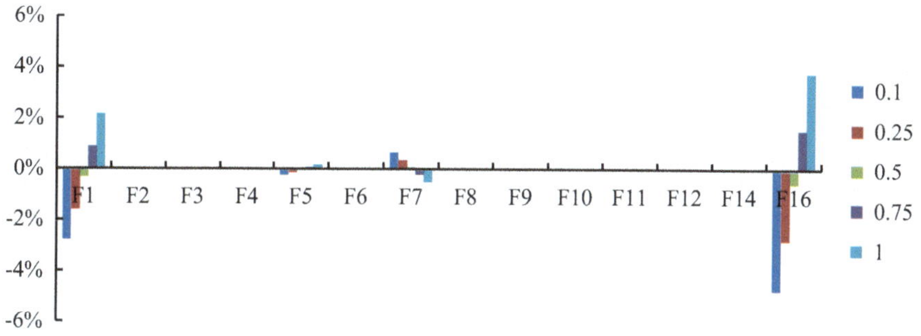

Fig. 7.6 System variation in the import security scenario

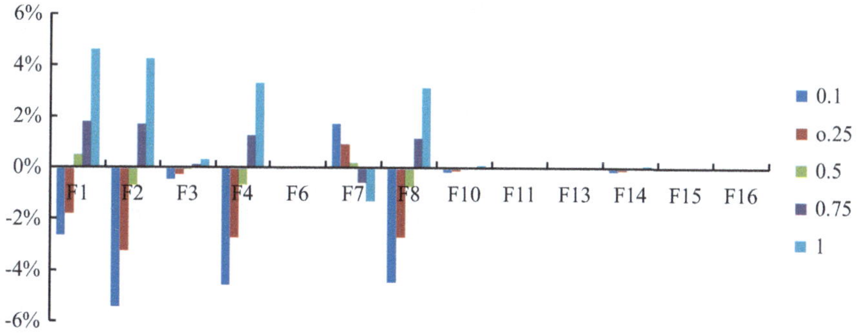

Fig. 7.7 System variation in the energy consumption reform scenario

case confirms that demand-side reform is helpful for the improvement of national energy security.

Moreover, another element concerning energy security is related to energy supply, and the results derived in the energy supply-side reform scenario are pictured in Fig. 7.8. It's clear that the impact of the energy supply-related factors on the system concentrates on the fluctuation in price stability (F7), which experienced a descent over 10% in the worst situation while a growth of more than 6% in the optimal case. Therefore, the energy security statue (F1) increases by close to 6% in the optimal case and declines by nearly 8% under the worst circumstance.

Besides, in the public participation scenario, a sensitivity analysis is conducted to highlight the role of the resident's awareness, and the corresponding results are presented in Fig. 7.9. It's worth noting that public awareness of the energy system influences many other elements in the system. Under the impact of the change of the corresponding factors, renewable energy production (F4) and import stability (F13) who vary from −8% to 8% are the most affected elements, followed by the environmental sustainability (F9) with a fluctuation from −5.9% to 4.3% and decentralization and electrification (F8) ranging between −5.1% and 4.1%. The results demonstrate that enhancing public awareness of energy policy and strategy can

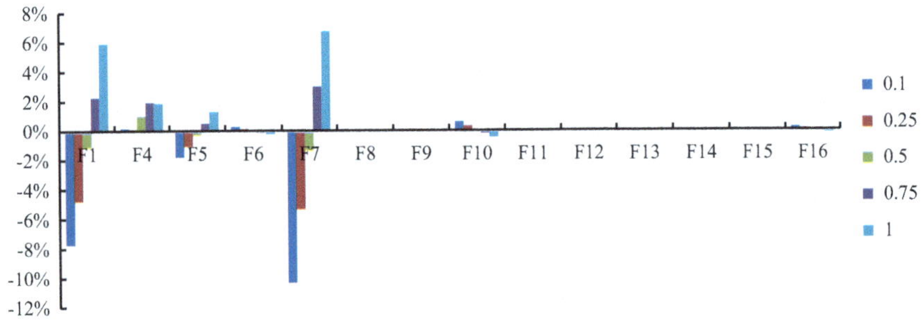

Fig. 7.8 System variation in the energy supply reform scenario

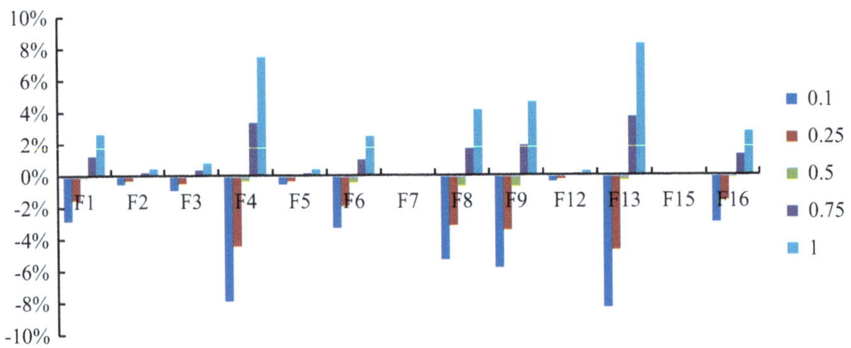

Fig. 7.9 System variation in the public participation scenario

help to improve residents' satisfaction with the current energy system to acknowledge the necessity of various energy development plans involving energy import, environmental issues, renewable energy, etc.

7.5 Conclusions and Policy Implications

Energy security plays a vital role in the social operation and economic development. However, the improvement and enhancement of national energy security confronts many obstacles, because the energy system involves a series of factors from various aspects, including supply, consumption, technological development, structure, and international climate and geopolitics, governance, as well as the residents' attitude to the development path. Through the construction of intuitionistic fuzzy cognitive map for energy security, this study aims to explain the cause-and-effect relationships between these factors and illustrate the influence mechanism by virtue of integrating intuitionistic fuzzy logic and fuzzy cognitive map to identify the role of various components concerning different energy strategies.

Firstly, for the improvement of energy security, reforms are needed at both the energy supply and consumption sides. There is a clear desire to increase the use of renewable energy to decrease dependence on foreign fuels, which has a positive effect on the improvement of China's energy security. However, this effect is only limited to the enhancement of energy diversity, and failed to improve energy security from other aspects. So, it's necessary to stress the integration of the green development plan with other energy strategies. What's more, China's energy security is threatened mainly due to its massive energy demand, which can hardly be met in a single way. Thus, effective management of energy demand is critical for enhancing China's energy security. The simulation results show that more attention should be paid to the research and development of new energy and environment-related technologies and the enhancement of industrial infrastructure to improve energy efficiency and achieve more environmental benefits for residents. Also, the energy-intensive industries need to be reformed and updated through industrial integration to improve efficiency.

As a country highly dependent on external energy, China needs to manage its energy imports. It's evident that energy import has a direct influence on national energy security. However, this effect is different with that from other factors. Instead, management of energy import is regarded as a relatively independent component in the energy system. In order to ensure energy import security, diversification is a traditional strategy. Presently, China has already established a relatively diversified energy import network, however, it still confronts threats from geopolitics. As countries in Middle East Asia has accounted for a considerable share both in global fossil energy reserves and China's energy imports, and these countries are also vulnerable to military and religious conflicts. Some other energy suppliers of China, such as Myanmar and Venezuela, have also been suffering from political turmoil and economic instability. Therefore, China should enhance its role in maintaining regional peace and stability with its international influence. The difficulties in conveying energy products from overseas to domestic is another barrier due to the high transportation costs and various human or natural induced risks in transportations. So more reliable energy transport modes and carriers should be established.

Besides, the results illustrate that public attitudes toward energy system are also taken as a crucial point for the improvement of energy security. Public attitudes have a fundamental impact on the other factors in the system, and people may hesitate to adopt new energy technologies because of unfamiliarity, particularly in remote regions. Thus, the formulation of the government's energy strategy and development plans should consider the effect of public acknowledgement, to motivate public enthusiasm and encourage their participation and understanding in the energy revolution. To do that, the development of energy policies, plans, and projects should consider the benefits of local citizens, disclose relevant information, and involve more motivations to get wide social participation and cooperation.

References

Amirkhani A, Mosavi MR, Mohammadi K, Papageorgiou EI (2016) A novel hybrid method based on fuzzy cognitive maps and fuzzy clustering algorithms for grading celiac disease. Neural Comput Appl 30(5):1573–1588

Atanassov KT (1986) Intuitionistic fuzzy sets. Fuzzy Sets Syst 20(1):87–96

Atanassov KT (1999) Intuitionistic fuzzy sets. Springer, pp 1–137

Axelrod R (1976) Structure of decision: the cognitive maps of political elites. Princeton University Press

Bekhrad K, Aslani A, Mazzuca-Sobczuk T (2019) Energy security in Andalusia: the role of renewable energy sources. Case studies in chemical and environmental engineering, p 100001

Blum H, Legey LF (2012) The challenging economics of energy security: ensuring energy benefits in support to sustainable development. Energy Econ 34(6):1982–1989

Bueno S, Salmeron JL (2009) Benchmarking main activation functions in fuzzy cognitive maps. Expert Syst Appl 36(3):5221–5229

Chen W, Yin X, Ma D (2014) A bottom-up analysis of China's iron and steel industrial energy consumption and CO2 emissions. Appl Energy 136:1174–1183

Cherp A, Jewell J (2014) The concept of energy security: beyond the four As. Energy Policy 75:415–421

Chester L (2010) Conceptualising energy security and making explicit its polysemic nature. Energy Policy 38(2):887–895

CPEA (2019) Blue book of china oil and gas industry development analysis and outlook report (2018–2019)

Criekemans D (2011) The geopolitics of renewable energy: different or similar to the geopolitics of conventional energy. In: ISA annual convention, pp 16–19

Dursun M, Gumus G (2020) Intuitionistic fuzzy cognitive map approach for the evaluation of supply chain configuration criteria. Math Methods Appl Sci 43(13):7788–7801

Gasser P (2020) A review on energy security indices to compare country performances. Energy Policy 139:111339

Gnansounou E (2008) Assessing the energy vulnerability: case of industrialised countries. Energy Policy 36(10):3734–3744

Groumpos PP (2010) Fuzzy cognitive maps: basic theories and their application to complex systems. Fuzzy cognitive maps. Springer, pp 1–22
Han Y, Deng Y (2018) A hybrid intelligent model for assessment of critical success factors in high-risk emergency system. J Ambient Intell Humaniz Comput 9(6):1933–1953
Hache E (2018) Do renewable energies improve energy security in the long run? Int Econ 156:127–135
IEA (2007) Energy security and climate policy-assessing interactions. OECD Publishing
Jacobsson S, Lauber V (2006) The politics and policy of energy system transformation—explaining the German diffusion of renewable energy technology. Energy Policy 34(3):256–276
Jewell J, Cherp A, Riahi K (2014) Energy security under de-carbonization scenarios: an assessment framework and evaluation under different technology and policy choices. Energy Policy 65:743–760
Klare MT (2006) The futile pursuits of energy security by military force. Brown J World Aff 13:139
Kosko B (1986) Fuzzy cognitive maps. Int J Man Mach Stud 24(1):65–75
Kruyt B, van Vuuren DP, de Vries HJM, Groenenberg H (2009) Indicators for energy security. Energy Policy 37(6):2166–2181
Lin B, Raza MY (2020) Analysis of energy security indicators and CO2 emissions. A case from a developing economy. Energy 200
Mendonça M, Lacey S, Hvelplund F (2009) Stability, participation and transparency in renewable energy policy: lessons from Denmark and the United States. Policy Soc 27(4):379–398
Novikau A (2019) Conceptualizing and achieving energy security: the case of Belarus. Energy Strat Rev 26
Papageorgiou EI, Markinos AT, Gemtos TA (2011) Fuzzy cognitive map based approach for predicting yield in cotton crop production as a basis for decision support system in precision agriculture application. Appl Soft Comput 11(4):3643–3657
Paravantis J, Kontoulis N, Ballis A, Tsirigotis D, Dourmas V (2018) A geopolitical review of definitions, dimensions and indicators of energy security. In: 2018 9th international conference on information, intelligence, systems and applications (IISA). IEEE, pp 1–8
Petroleum B (2020) BP statistical review of world energy
Ren J, Sovacool BK (2014) Quantifying, measuring, and strategizing energy security: determining the most meaningful dimensions and metrics. Energy 76:838–849
Sovacool BK (2014) Environmental issues, climate changes, and energy security in developing Asia. In: Asian development bank economics working paper series, vol 399, pp 17–14
Sovacool BK, Mukherjee I (2011) Conceptualizing and measuring energy security: a synthesized approach. Energy 36(8):5343–5355
Vivoda V (2009) Diversification of oil import sources and energy security: a key strategy or an elusive objective? Energy Policy 37(11):4615–4623
Wüstenhagen R, Wolsink M, Bürer MJ (2007) Social acceptance of renewable energy innovation: an introduction to the concept. Energy Policy 35(5):2683–2691
Yao L, Chang Y (2014) Energy security in China: a quantitative analysis and policy implications. Energy Policy 67:595–604
Yao L, Chang Y (2015) Shaping China's energy security: the impact of domestic reforms. Energy Policy 77:131–139
Zadeh LA (1965) Fuzzy Sets. Inf Control 8(3):338–353
Zeng EY, You J, Cheng H (2011) Balance between economic growth and environmental protection: sustainability through better science. J Environ Monit 13(4):787–788
Zha D (2006) China's energy security: domestic and international issues. Survival 48(1):179–190
Zhou P, Ang BW (2008) Linear programming models for measuring economy-wide energy efficiency performance. Energy Policy 36(8):2911–2916

Chapter 8
Quantitative Risk Assessment and Management of Cash-In-Transit Vehicle Routing Problems

Yuanzhi Jin, Xianlong Ge, and Long Zhang

Abstract Safe and scheduled money and bill transportation is critical for improving the efficiency of banking industry, while various risk factors have affected the daily pick-up and delivery activities of cash-in-transit (CIT) sectors. Thus, it is necessary to develop a model to find the most reliable route for CIT sectors which can efficiently complete the pick-up and delivery activities with the minimum risks. Here this paper investigates the existing research efforts and conducted a comprehensive overview of these works on risk assessment in the CIT industry. In this paper, we summarized and divided the methods of risk assessment into three categories according to the calculations on routes. In addition, we also present a mathematical model for cash-in-transit vehicle routing problems (CTVRP) considering risk constraints and some solution algorithms for the model. Finally, a comprehensive analysis of risk management for CIT sectors is discussed. To our best knowledge, this paper might help researchers get a comprehensive understanding of related research in the cash-in-transit sectors.

Keywords Risk assessment · Cash-in-transit · Vehicle routing problem · Risk constraints · Solution algorithms

8.1 Introduction

The banking industry plays a critical role in providing financial support for economic and social development. In order to further stimulate market potential and help the development of small and medium-sized enterprises, financial inclusion has been

Y. Jin · X. Ge (✉)
School of Economics and Management, Chongqing Jiaotong University, Chongqing 400074, China
e-mail: gexianlong@cqjtu.edu.cn

Y. Jin
Department of Computer Technology and Information Engineering, Sanmenxia Polytechnic, Sanmenxia 472000, China

L. Zhang
School of Business, Xinyang Normal University, Xinyang 464000, China

J. Ren (ed.), *Multi-Criteria Decision Analysis for Risk Assessment and Management*, Industrial Ecology and Environmental Management 1,
https://doi.org/10.1007/978-3-030-78152-1_8

provided. Financial inclusion indicates that people or businesses can visit important and necessary financial products and services that meet their finance-related needs (IBRD 2018). Great strides have been made toward financial inclusion, and 1.2 billion adults worldwide have gotten access to an account since 2011. Although 69% of adults have an account today, close to one-third of adults are still unbanked. About half of unbanked people are living in less developed areas or unemployment (IBRD 2018). Financial inclusion is the key to reduce poverty and boost prosperity because it can improve the overall quality of people's life.

In order to improve the efficiency of the banking system, efficient cash pick-up and delivery activities are essential. The optimal cash pick-up or delivery is beneficial to reducing the cost of cash transportation in the banking industry, which is of great significance for promoting the rapid development of financial inclusion. Generating efficient routes for cash pick-up or delivery belongs to cash-in-transit vehicle routing problems, which is rooted in the classical vehicle routing problem (VRP). Since Dantzig and Ramser (1959) proposed the VRPs, a large body of literature has generated various models related to it, and the capacitated vehicle routing problem (CVRP) is of the variants. In order to solve CVRPs, many exact algorithms have been proposed and developed, such as branch-and-cut algorithm (Letchford et al. 2007) and branch-and-cut-and-price algorithm (Santos et al. 2015). However, exact algorithms can only find the solutions for small-scale instances, and many large-scale cases in practical applications cannot be solved in limited time with exact algorithms. In order to deal with this problem, scholars have developed a large number of heuristic or meta-heuristic algorithms for the CVRP, such as savings method (Stanojević et al. 2013), sweep algorithm (Gillett and Miller 1974), ant colony optimization (ACO) (Bell and McMullen 2004), and genetic algorithm (GA) (Lin et al. 2019). These heuristic algorithms have shown very good performance for solving large-scale CVRPs.

To improve the practicability and applicability of the CVRP model, scholars have considered an increasing number of factors or constraints. For example, the CVRP model with consideration of risk factors related to the routes is also applicable for solving problems of transportation valuable goods such as cash, jewelry, and antiques. The CVRP model is transformed into CTVRP if cash is transported. In fact, most studies on transportation risks in CVRP problems are concerning the routing of hazardous materials, while very little attention has been paid on routing problems in the cash-in-transit industry. There are various mathematical models that have been used in hazmat transportation, which capture different risk aspects in transportation (Pradhananga et al. 2010, 2014a, b; Kazantzi et al. 2011; Androutsopoulos and Zografos 2012; Parsafard et al. 2015; Chen et al. 2017; Kumar et al. 2018; Bula et al. 2019; Ghaderi and Burdett 2019). Generally, risk functions are defined according to the road conditions and characteristics of transported dangerous goods. For CIT industry, population surrounding the routes is not endangered like that in the transportation of hazardous materials, and there are also no risks or probability of explosion. Instead, the vehicles are under the risk of robbery along their routes.

The existing literature focusing on cash-in-transit vehicle routing problems can be divided into three categories according to the different methods of risk calculation.

(1) Distance-based methods. These risk assessment methods are based on the distance the vehicles travel (or the duration the vehicles expose) along the arcs of a route. Generally, the risk of the CIT vehicle being robbed is positively related to the travel distance (or duration) along the routes. The greater the distance the vehicle travels, the higher the risk of robbery is.
(2) Probability-based methods. The conduction of these risk assessment methods depends on several factors such as the condition of a road segment, social structure of a region, usage frequency of the paths in multiple horizons, and weather conditions. The factors are subsumed to evaluate the probability of robbery and loss after a successful robbery.
(3) Dissimilarity-based methods. Theoretically, the probability of being robbed is very small if the routes cannot be predicted in time and space during the cash pick-up or delivery process. Therefore, these risk assessment methods try to generate peripatetic routes in the successive multiple horizon, making the vehicle more unpredictable. The randomness of peripatetic is measured by the dissimilarity between the solutions in multiple successive horizons.

The residual parts of this paper are organized as follows: In Sect. 8.2, the CIT business and CTVRP are presented. In Sect. 8.3, the three kinds of methods of risk assessment are reviewed and the other related methods are introduced. In Sect. 8.4, we introduce the CTVRP model with risk constraints. Meanwhile, the solution algorithms of CTVRP are described briefly. Section 8.5 includes a discussion of risk assessment methods. Finally, this study has been concluded in Sect. 8.6.

8.2 Problem Description

8.2.1 CIT Business in China

CIT sectors usually provide cash transport services between branches and a central bank, and they need to deliver or pick-up cash timely for each branch with a limited number of armored vehicles every day. The vehicles deliver cash box pairs (including two boxes, one for cash and another for bills in China) from a central bank to each branch in the morning, and the pick-up procedure is reversed in the evening. The cash containers to be transported from or to each branch are marked by code, and the cash and bills demanded for each branch can be held in only one box pairs per day. If a branch needs a large amount of cash, an appointment is required in advance. Additionally, each vehicle can hold up to 30 boxes, i.e., serve at most 15 branches.

8.2.2 *Description of CTVRP*

Generally, the CTVRP can be defined on a directed or undirected graph $G = (V, A)$ (Talarico et al. 2015a). There is a central bank (or depot, numbered 1) where all CIT vehicles depart from and return to. Then, the nodes' set $V = \{1\} \cup C$ corresponds to the customers' set $C = \{2, 3, \ldots, n + 1\}$ and the central bank. Each customer $i \in N$ has a demand q_i of cash pick-up or delivery by a vehicle during its visit. Specially, q_i should be considered as unit demand because each branch only needs one box pairs per horizon in China. A nonnegative distance d_{ij} is associated with each $arc(i, j) \in A$. All vehicles depart from the central bank and carry out a single route. Each vehicle has to visit a set of ordered customers before returning to the central bank. Assuming that each vehicle's risk index is equal to zero at the central bank during the cash pick-up process. A vehicle traveling on the $arc(i, j)$ increases its risk value that can be calculated by formula (8.3). Similarly, each vehicle's risk value reaches to a maximum along the route assigned when it departs from the central bank during the cash delivery process. A vehicle traveling on a router r decreases its risk value because the on-board cash is gradually declined when it leaves the last visited node (See Fig. 8.1).

8.3 Methods of Risk Assessment

According to Sect. 8.1, the methods for assessing risks of cash-in-transit vehicle routes can be divided into three types according to the calculation of risk indexes. After introducing the origin and model of VRPs in Sect. 8.2, we present a detailed description of three kinds of methods of risk assessment proposed in the literature here.

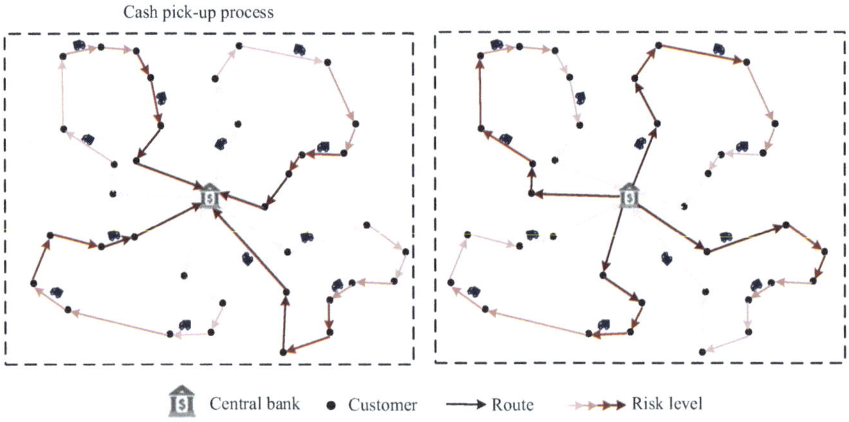

Fig. 8.1 An illustration of CTVRPs with risk assessment

8.3.1 Distance-Based Methods

The definition of risk mainly consists of three parts: (a) the occurrence of an incident (expressed as the probability or frequency of a particular unwanted incident happening); (b) the vulnerability as a measure of the susceptibility of a protected object to harmful events; (c) the exposure that denotes a weighted value of the people, protected object, and infrastructure affected during and after the incident (Russo and Rindone 2011). Following this concept, a route r may include many arcs ($\geq$2) and the risk of being robbed in any *arc*(*i*, *j*) can be calculated by formula (8.1) (Talarico et al. 2015b).

$$R_j^r = p_{ij} \cdot v_{ij} \cdot D_i^r, \tag{8.1}$$

where p_{ij} represents the probability of a robbery happening, v_{ij} is a measure of the probability that the robbery succeeds given that it occurs, and D_i^r means the quantifiable losses after the robbery succeeds. Assuming that the number of robberies on route r is no more than 1, the aggregated risk along route r can be calculated by formula (8.2)

$$GR^r = \sum_{arc(i,j) \in r} p_{ij} \cdot v_{ij} \cdot D_i^r. \tag{8.2}$$

However, it is very difficult to determine the values of p_{ij} and v_{ij} in practical applications. Therefore, the authors (Talarico et al. 2015a) assumed that the probability is proportional to the length of *arc*(*i*, *j*) and that v_{ij} is to be a constant for all arcs, which can be calculated by using the past records of successful attacks in the target area. Finally, to determine the route risk during the pick-up process, a risk index R_j^r is computed for each node i in route r in a recursive manner (Talarico et al. 2015b) as follows:

$$R_j^r = R_i^r + d_{ij} M_i^r, \tag{8.3}$$

where R_j^r represents a cumulative risk of the route r, M_i^r denotes the on-board money of the vehicle when it leaves node i to node j, and d_{ij} is the length of *arc*(*i*, *j*) included in r. For the sake of simplicity, the constant v_{ij} is omitted.

The example in Fig. 8.2 could interpret the risk assessment for the route r. The values of weight on the arcs represent their travel distance or duration, and each customer has a demand q_i, $i \in \{A, B, C\}$. Assumed that the visit order of a vehicle is 1- > A- > B- > C- > 1. The risk index $R_A = 0$ because the vehicle travels along arc (1, A) without load (see Fig. 8.2a). At the second node the vehicle pick-ups two units of cash and continues traveling to the next node. According to the formula (8.3), the risk at node B can be calculated (see Fig. 8.2b). Similarly, the value of risk at node C is equal to 45 (see Fig. 8.2c). Finally, the global risk of the route r is equal to 69

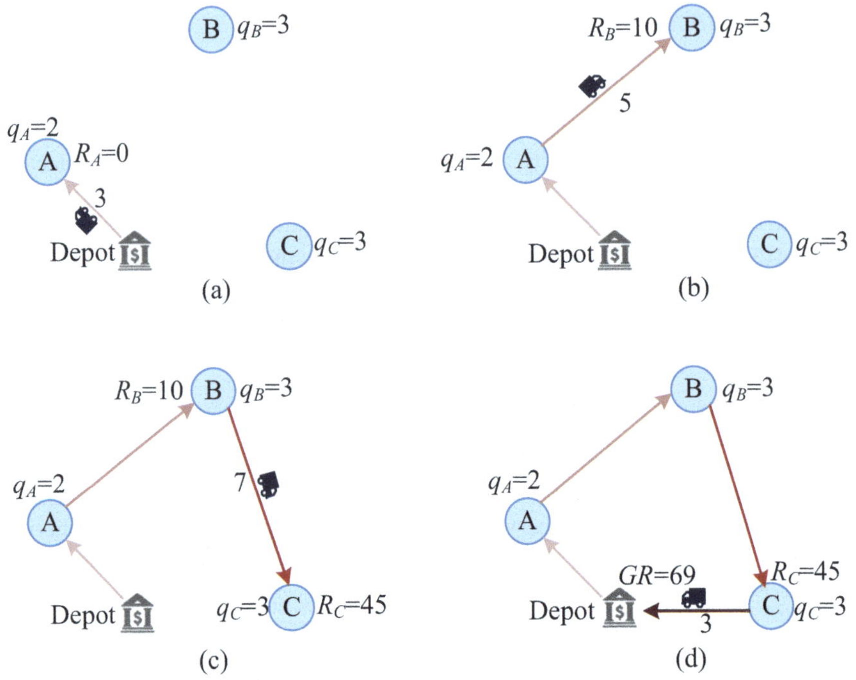

Fig. 8.2 An example of risk assessment for a route during cash pick-up process

when the vehicle travels on the arc (C, 1) (see Fig. 8.2d). During the optimization process, the route r is safe enough if the global risk is at most equal to a predefined threshold θ.

The calculation method of the risk index during delivery process is different from formula (8.3) because the delivery risk is slowly released along the routes. Similarly, the delivery risk of the route r is defined as follows:

$$R_i^r = d_{i,i+1} M_i^r, \tag{8.4}$$

where $d_{i,i+1}$ refers to the distance from node i to the next node. Generally speaking, there are less risks in the delivery process than the pick-up. In other words, as long as the maximum risk of routes during the cash pick-up process does not exceed a threshold value θ, the corresponding risk during the delivery process can be ignored, which may be the reason why the authors (Talarico et al. 2015a) did not give formula 8.4. Based on this risk assessment method, many scholars have done meaningful research and even made some modifications and improvements (Xu et al. 2019).

Using the same model and risk assessment methods, Talarico et al. (2017a) proposed a new heuristic algorithm, called Ant Colony Optimization with Large Neighborhood Search (ACO-LNS), to solve the Risk-constrained Cash-in-Transit Vehicle Routing Problem (RCTVRP). At the same time, the similar authors (Talarico

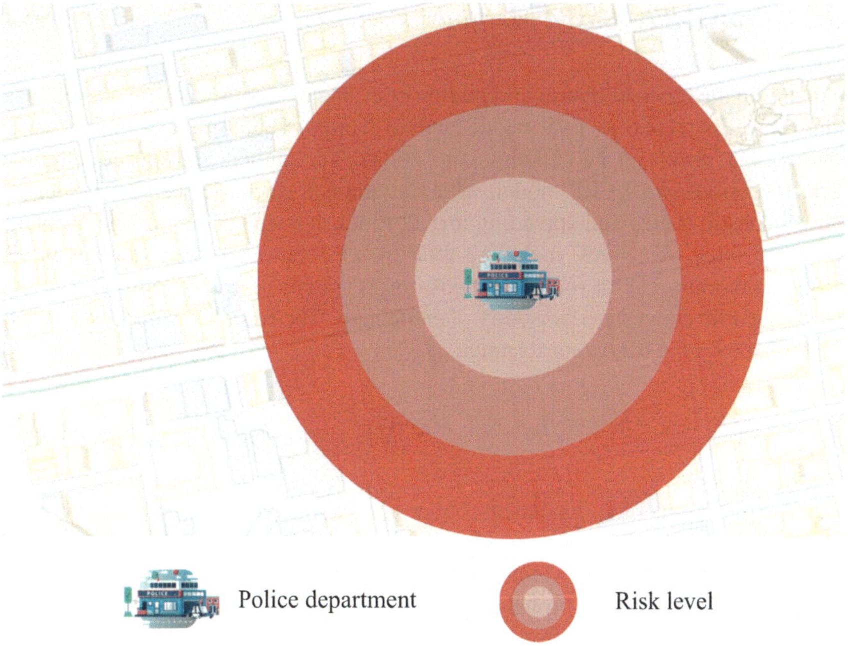

Fig. 8.3 An illustration of risk levels around a police department

et al. 2017b) considered the max risk of routes as the second objective and designed a Progressive Multi-Objective Optimization with Iterative Local Search (PMOO-ILS) to deal with the Multi-Objective RCVRP (MO-RCVRP).

Note that, the predefined threshold θ in the model from (Talarico et al. 2015a) is considered as hard constrains, which are satisfied if the global risk index is not larger than θ for each route, However, these constraints do not take into account how much smaller than θ the risk indexes are. In order to solve this tension, Radojičić et al. (2018a) proposed a new fuzzy variant of the RCTVRP for more adequate modeling of safe routes using fuzzy numbers based on the threshold constant θ. The similar method was employed in the authors' other two papers (Radojičić et al. 2018b; Radojičić and Marić et al. 2018).

A risk assessment method similar to (Talarico et al. 2015b) has been adopted to solve the multi-horizon RCTVPR (Guerrero et al. 2020). The authors proposed a model of Risk Constrained Inventory Routing Problems with Time Windows (RCIRPTW) in bank correspondents, and the risk constrains are modified to fit the multi-horizon problems.

8.3.2 Probability-Based Methods

No matter in the process of cash distribution or cash pick-up, the risk of CIT vehicles being robbed is related to a variety of random factors. Therefore, a comprehensive framework is proposed by (Tikani et al. 2021). The authors introduced a multi-objective periodic RCTVRP model, which attempts to increase security by generating stochastic routes and spreading arrival times at each customer. The problem minimized three objectives, including total travel durations for all vehicles in all periods, maximum risk of routes, and customers' satisfaction level considering the effects of traffic congestion as a daily phenomenon. The risk along *link* (i, j, m) can be calculated by the following formula:

$$R_{ijm}^{p} = \mathrm{I}_{ijm}^{p} \cdot v_{ijm}^{p} \cdot D_{ik}^{p} \cdot \left(t_{jk}^{p} - t_{ik}^{p} - s_j\right), \tag{8.5}$$

where I_{ijm}^{p} represents time-dependent robbery probability per time unit, which relies on several factors such as the condition of a link, social structure of a region, usage frequency, and weather conditions. v_{ijm}^{p} and D_{ik}^{p} are similar to v_{ij} and D_i^r in formula (8.2), indicating the probability that the robbery succeeds given that it occurs and the quantifiable losses after the robbery succeeds, respectively. The last part $\left(t_{jk}^{p} - t_{ik}^{p} - s_j\right)$ in formula (8.5) means the travel time along the *link* (i, j, m) associated with the route k. Since it is very difficult to estimate the probability of robbery in different target areas, the parameter v_{ijm}^{p} is ignored. Meanwhile, in the first period, the robbery probability of each link I_{ijm}^{p} may be set to a value in uniform distribution of [0, 0.2] randomly. And then in the subsequent periods, the robbery probabilities are computed using recursive formula (8.6).

$$\mathrm{I}_{ijm}^{p} = \mathrm{I}_{ijm}^{p-1} \cdot \left(1 + \psi_{ijm} \sum_{k \in K} \sum_{h \in H_{mij}} x_{ijm}^{hkp-1}\right), \tag{8.6}$$

where ψ_{ijm} is an intensity factor related to the usage frequency of links during different periods.

In order to generate variant safe routes, the arrival time at each customer node must be spread during the planning horizon (Hoogeboom and Dullaert 2019). Therefore, the new arrival-time diversification constraints (8.7) were added to their model (Soriano et al. 2020).

$$\left|t_{ik}^{p} - t_{ik}^{p'}\right| \geq \xi_i \forall i \in N, \forall p, p' \in P, p \neq p', \tag{8.7}$$

where the parameter ξ_i is a minimum time–space variability for customer i that is initialized based on experts' poll. Finally, based on the formula (8.3), the total risk of route k after passing node i through node j by link m in period p is defined as follows:

$$R_{jk}^{p} = R_{ik}^{p} + \mathrm{I}_{ijm}^{p} \cdot D_{ik}^{p} \cdot \left(t_{jk}^{p} - t_{ik}^{p} - s_{j}\right). \tag{8.8}$$

Moreover, the same authors of (Tikani et al. 2021) dealt with the similar problems by using the framework of risk assessment (Tikani et al. 2020). The new efficient algorithms, however, including a novel flexible restricted dynamic programming and a modified genetic algorithm are designed to deal with the multi-objective periodic RCTVRP.

The estimation of the robbery probability p_{ij} and the probability of successful robbery v_{ij} has always been up in the air. Ghannadpour and Zandiyeh (2020) utilized the method based on game theory to calculate the theft probabilityp_{ij} and the multiple-criteria decision-making (MCDM) to evaluate the probability v_{ij} of theft success. In order to calculate p_{ij}, the linear scheduling model of game theory from the attacker's viewpoint was formulated. As for the estimation of v_{ij}, the three attributes, including the number of police stations, the specification of a street, and rate of crimes, were taken into account during the decision process. The detailed information about this method of risk assessment can be found on pages 2–10 of (Ghannadpour and Zandiyeh 2020).

8.3.3 Dissimilarity-Based Methods

Combined with a perfect b-matching algorithm, Ngueveu et al. (2009) proposed a framework of the tabu search for m-Peripatetic Vehicle Routing Problems (m-PVRPs). For security reasons, the framework ensured that no path can be used more than once during m consecutive periods, which may reduce the chance of a potential robbery. The more random the path is, the less likely CIT vehicles are to be robbed.

To make it more difficult to track vehicles, thereby reducing the risk of robbery, there must be variation in daily vehicle routes and schedules, including the daily vehicle arrival time at each demand point (Yan et al. 2012). The RCTVRP was formulated as an integer multiple-commodity network flow problem, which incorporates a new concept of similarity of time and space for routing and scheduling to reduce the risk of robbery (Yan et al. 2012). Firstly, the variation in vehicle arrival time at each demand point is evaluated, and then the tour similarity can be calculated to evaluate the similarity of time and space for each vehicle tour. Two years later, the same authors established a stochastic cash transportation vehicle route and schedule model by using the concept of time–space similarity (Yan et al. 2014).

Although the above approach is capable of assessing the similarity of different vehicle routes and schedules, it is not applicable to scenarios that fail to consider time windows. More generically, in order to quantify the dissimilarity between two paths P_i and P_j, Martí et al. (2009) proposed a new method based on spatial Euclidean distance as follows:

$$Dis(P_i, P_j) = \frac{1}{2}\left[\frac{\sum_{v_l \in P_i} \Phi(v_l, P_j)}{|P_i|} + \frac{\sum_{v_l \in P_j} \Phi(v_l, P_i)}{|P_j|}\right], \tag{8.9}$$

where the value $\Phi(v, P_j)$ represents the Euclidean distance from a vertex v to a path $P_j = \{v_1, v_2, \ldots, v_n\}$, and it is expressed as follows:

$$\Phi(v, P_j) = \min_{v_l \in P_j} \Phi(v, v_l) \tag{8.10}$$

Another method measuring the dissimilarity between two paths P_i and P_j based on the length of shared edges is designed by (Akgün et al. 2000) and (Vanhove 2012). The formula to compute the dissimilarity is as follows:

$$Dis(P_i, P_j) = 1 - \frac{1}{2}\left[\frac{L(P_i \cap P_j)}{L(P_i)} + \frac{L(P_i \cap P_j)}{L(P_j)}\right], \tag{8.11}$$

where $L(*)$ is a function that calculates the length of the routes or segments.

In order to obtain a set of solutions with high dissimilarity for safe route planning and scheduling, an additional mathematical formulation of the k-dissimilar vehicle routing problem (*kd*-VRP) is introduced as follows (Talarico et al. 2015a):

$$\min \max_{i \in \{1,2,\ldots,k\}} f(y_i) \tag{8.12}$$

s.t.

$$\delta(y_i, y_j) \le \mathrm{Ts}, \forall i, j \in \{1, 2, \ldots, k\}, i \neq j, \tag{8.13}$$

$$y_i \in \Omega, \forall i \in \{1, 2, \ldots, k\}. \tag{8.14}$$

The objective function (8.12) minimizes the cost of the worst alternative solution in the solution set resulted from the *kd*-VRP model. Ts represents the maximum similarity between alternative solutions, and it is a parameter of the problem that should be defined as an input by the user. A new function $\delta(y_i, y_j)$ for measuring the dissimilarity between two paths P_i and P_j based on formula (8.11) is defined as follows Talarico et al. (2015a):

$$\delta(y_i, y_j) = \max_{l,m \in N} \frac{1}{2}\left[\frac{Cs\left(r^l_{y_i}, r^m_{y_j}\right)}{Cs\left(r^l_{y_i}\right)} + \frac{Cs\left(r^l_{y_i}, r^m_{y_j}\right)}{Cs\left(r^m_{y_j}\right)}\right]. \tag{8.15}$$

The set N includes identical vehicles, each with a limited capacity. The function Cs with one parameter means calculating the corresponding cost of a route and with two parameters represents the aggregate cost of the edges shared between routes.

According to formula (8.15), the similarity between two solutions $\delta(y_i, y_j)$ is a value in the range from 0 to 1, which means that the parameter Ts should be set in the interval of [0, 1]. The k solutions with any shared edge are forbidden if $Ts = 0$. On the right counterpart, the same solutions may be generated.

8.3.4 Other Methods

The distance between the CIT vehicles and the police stations is also an effective indicator for measuring route risks of CIT vehicles, and the shorter distance indicates safer vehicle routes (see Fig. 8.3). Therefore, contours of different risk levels can be determined around the police departments along the CIT routes. Tarantilis and Kiranoudis (2004) developed a decision support system (DSS) for logistics planners of a well-known hank to design safe delivery routes. The solutions obtained by DSS focus on minimizing the probability of successful vehicle robbery at a certain point of the road network. The risk is measured by the distance between the risk-point and the closest police department.

Both academic researchers and law enforcement agencies commonly use terms such as hot spots, hot routes, or crime clusters referring to spatial concentration of crime. Neighborhoods with a low socio-economic status (SES), e.g., those with a high level of poverty and unemployment rate, are more likely to be hot spots of offences (Geyer 2007).

Moreover, some studies used different approaches from the above-mentioned equations. For instance, Bozkaya et al. (2017) proposed a new model of risk assessment, namely, usage-based risk index (UBRI), for transport of valuable goods in which two factors of SES and the usage frequency of links were employed. The UBRI value of a particular path π_{ijk} on any given day d was calculated as formula (8.16).

$$UBRI_{ijkd} = \sum_{e \in \pi_{ijk}} f_{ed}, \tag{8.16}$$

where

$$f_{ed} = \begin{cases} \alpha \cdot n_{e,d-1} + (1-\alpha) f_{e,d-1}, \; if \; under \; the \; initial \; stages \\ (f_{ed} - \mu), \; if \; f_{ed} > \mu \; and \; after \; some \; initial \; stages \\ 0, \; otherwise \end{cases} \tag{8.17}$$

The SES values of the underlying geographical region were employed to calculate a normalized socio-economic risk index (SERI) of vertices and arcs (Bozkaya et al.

2017). And the SERI value of the *k-th* path π_{ijk}, $k \in \prod_{ij}$ from vertex *i* to vertex *j* is calculated as formula (8.18).

$$SERI_{ijk} = \sum_{(u,v)\in\pi_{ijk}} l_{uv} \cdot \frac{SERI_u + SERI_v}{2}, \tag{8.18}$$

where $SERI_u$ and $SERI_v$ refer to the *SERI* values of the neighborhoods where vertices *u* and *v* reside, which can be estimated by the surrounding population in advance.

Finally, the mentioned equation was calculated as formula (8.19).

$$R_{ijkd} = w_U c_U UBRI_{ijkd} + w_S c_S SBRI_{ijk}, \tag{8.19}$$

where R_{ijkd} is the risk of link from vertex *i* to vertex *j* in day *d* with vehicle *k*. w_U and w_S are the weights associated with these two risk elements. c_U and c_S are also scaling coefficients for the two elements of the cost function. Therefore, the resulting composite cost of a path π_{ijkd} on day *d* is defined in formula (8.20).

$$C_{ijkd} = w_D c_D d_{ijk} + w_R R_{ijkd}. \tag{8.20}$$

Here, the constants c_D are scaling coefficients for the three elements of the cost function. w_D and w_R are the unified weights of the two risk measures, which have a relationship of $w_D + w_R = 1$.

Hepenstal and Johnson (2010) discovered that CIT robberies were prone to occur in the hotspots of cities based on the analysis of the historical data of crime. The findings suggest that CIT robbery clustered in space more than would be expected on the basis of the distribution of targets and that the risk of CIT robbery was particularly acute around major intersections. Although vehicles will inevitably pass through some hot spots and major traffic intersections in urban areas, the schedules of CIT vehicles can try their best to pass through these areas as little as possible during the route planning stage (especially the routes between two adjacent nodes) to reduce the risk of robbery based on these findings. In fact, these robbery-prone areas have a common feature: heavy traffic. Vehicles traveling in these areas tend to run at slower speeds, which increase the likelihood of being robbed. Therefore, the paths with higher travel speed should be selected as the first choice in the route planning stage, even though it may increase the travel distance.

8.4 A Demo Model of CTVRP

The CTVRP model studied in this paper belongs to a special VRPTW model, which has a characteristic that each bank branch has the same demand and the uniform time window. The over-concentrated time windows bring great challenges to the efficient dispatching of vehicles. In order to tackle this problem, we have designed a CTVRP model and introduced algorithms for solving it.

8.4.1 The CTVRP Model with Risk Constraints

In this section, we establish a mathematical model for the CTVRP uniform time window. To introduce the model more conveniently, some nomenclatures are given first, as listed in Table 8.1.

Based on the information in Table 8.1 and the previous description, a CTVRP model that aims to minimize the total operational cost related to the vehicles is established.

$$Min\ z = V_c \cdot \sum_{k\in K} \sum_{j\in V} \sum_{i\in V} d_{i,j} \cdot x_{i,j}^k + \sum_{k\in K} F_c \cdot \max_{i\in c}(y_i^k) \tag{8.21}$$

Table 8.1 Nomenclature used in this paper

Nomenclature	Meaning
$C = \{2, 3, \ldots, n+1\}$	The set of customers
$V = \{1\} \cup C$	The set of nodes
$K = \{1, 2, \ldots, k\}$	The set of vehicles
M or M'	Sufficiently large positive number
q_i	The demand of customer *i*
w_i	The service duration of customer *i*
$t_{i,j}$	The travel time from node *i* to *j*
$d_{i,j}$	The travel distance from node *i* to *j*
T	The efficient working time
Q	The capacity of vehicles
D	The amount of onboard cash
F_c	The fixed cost of vehicles
V_c	The variable cost of vehicles
$\underline{U}$	The minimum number of customers that the vehicle needs to serve
S_f	The average service fees paid by customers each year
EY_c	The average total cost of vehicle *k* per year
α	The profit factor of the vehicle
m	The maximum number of vehicles available
θ	The risk threshold
$x_{i,j}^k$	1 if vehicle k travels the *arc* (*i*, *j*); 0 otherwise
y_i^k	1 if customer *i* is served by vehicle *k*; 0 otherwise

s.t.

$$\forall i \in C, k \in K: \sum_{j \in V} x_{i,j}^k = \sum_{j \in V} x_{j,i}^k = y_i^k \tag{8.22}$$

$$\forall i \in C: \sum_{k \in K} y_i^k = 1 \tag{8.23}$$

$$\underline{U} = \alpha \lceil EY_c / S_f \rceil \tag{8.24}$$

$$\forall k \in K: \sum_{i \in C} y_i^k \geq \underline{U} \tag{8.25}$$

$$\forall k \in K: \sum_{i \in C} q_i \cdot y_i^k \leq Q \tag{8.26}$$

$$\forall k \in K: \sum_{i \in C} \sum_{j \in C} t_{i,j} \cdot x_{i,j}^k + \sum_{i \in C} w_i \cdot y_i^k \leq T \tag{8.27}$$

$$\sum_{k \in K} \max_{j \in C} (y_j^k) \leq m \tag{8.28}$$

$$\forall k \in K: \sum_{j \in C} x_{1,j}^k = \sum_{i \in C} x_{i,1}^k \tag{8.29}$$

$$\forall H \subseteq C, \ h \in H, \ k \in K: \sum_{i \in H} \sum_{j \in \{V \backslash H\}} x_{i,j}^k \geq y_h^k \tag{8.30}$$

$$\forall k \in K: D_{1,k} = \sum_{i \in N} q_i \cdot y_i^k \leq Q \tag{8.31}$$

$$\forall k \in K, \operatorname{arc}(i,j) \in A: D_{j,k} \geq D_{i,k} - q_j y_j^k - M(1 - x_{i,j}^k) \tag{8.32}$$

$$\forall k \in K, \operatorname{arc}(i,j) \in A: R_i^k \geq D_{i,k} d_{i,j} - M'(1 - x_{i,j}^k) \tag{8.33}$$

$$\forall k \in K: \sum_{arc(i,j) \in A} R_i^k x_{i,j}^k \leq \theta \tag{8.34}$$

The objective z (8.21) is minimizing the total operational cost. Constraints (8.22) and (8.23) ensure that each customer is served only once on a horizon. Constraints (8.24) and (8.25) represent the minimum number of customers served per vehicle. Constraints (8.26) define the capacity constraint of the vehicles. Constraints (8.27) indicate that the vehicles arrive at customers within their time windows. Constraints (8.28) mean that the number of vehicles used cannot exceed the number of the fleet owned. Constraints (8.29) denote that all vehicles must start from and return to the central bank. Finally, sub-tours are eliminated by constraints (8.30). Constraints (8.31) ensure that the vehicle is not overloaded when leaving the depot. Constraints (8.32) represent the change in onboard cash of vehicle k as it passes through *arc*(i, j). Constraints (8.33) represent the risk constraint when vehicle k passes through *arc*(i, j). Finally, constraints (8.34) ensure that the global route risk is lower or equal to the risk threshold θ.

In order to verify the closure of the CTVRP model, the CPLEX solver (academic version: 12.10) can be employed to test its effectiveness. Performing on the randomly generated small-scale instances (n < 18), the optimal solution can be obtained in a few seconds; however, the solver becomes powerless when facing the larger instances (n > 18) because constraints (8.30) contain an exponential operation to enumerate all subsets of the set C. Therefore, heuristics are needed to solve large-scale instances in a reasonable time.

The method of risk assessment in the CTVRP model belongs to distance-based methods. Moreover, there are four practical ways to reduce risk during the routing optimization process.

(1) Vehicles can visit police stations nearby the route they service if possible.
(2) Try to pass through as few large intersections with traffic lights as possible.
(3) There are multiple routes between two nodes to choose under the developed urban traffic networks. Choosing as many routes as possible that travel faster so as to reduce the exposure time of CIT vehicles.
(4) Trying to avoid underdeveloped areas, such as "villages" in cities, slums, etc.

8.4.2 Solution Algorithms of CTVRP

Adding a risk threshold to the traditional VRP model as constraints for each route is a simple way to generate safe solutions because existing heuristics can be used directly to solve the new model without major modification. For example, if we want to solve CTVRP with a genetic algorithm that can solve the traditional VRP (Karakatič and Podgorelec 2015), we only need to determine whether the constraints (8.34) hold when a new chromosome is generated in each generation. If the chromosome representing a solution satisfies the constraints (8.34), it is retained; otherwise, a new chromosome is regenerated until the number of feasible chromosomes reaches the pre-set population size.

Another approach to dealing with risk constraints is to minimize the maximum route risk as an additional objective function (8.35).

$$\min z' = \max_{i \in C, k \in K} R_i^k \tag{8.35}$$

At this point, the model becomes a multi-objective one. At present, NSGA-II can be directly used to solve multi-objective VRP (Jemai et al. 2012). The algorithm is suitable for complex and multi-objective optimization problems, conquering the main defects, high time complexity of non-dominated sorting in NSGA when the population size is large, and achieving fast and accurate search performance.

Multi-objective problems can be transformed into a single objective function in advance if we try to use traditional heuristic algorithms designed for single-objective VRPs.

Assume that $\mathbb{R}^*$ represents * dimensional Euclidean vector space, $\mathbb{R}^n$ and $\mathbb{R}^p$ represent decision vector space and objective vector space, respectively. $X \subset \mathbb{R}^n$ is a feasible solution set, the objective function f: $\mathbb{R}^n \rightarrow \mathbb{R}^p$ composed of p real valued objective functions, i.e., $f = (f_1, f_2, \ldots, f_p)$ where f_k:$\mathbb{R}^n \rightarrow \mathbb{R}, k = 1, 2, \ldots, p$. Therefore, a Multi-objective Optimization Problem (MOOP) can be defined as follows:

$$
\begin{gathered}
Min(f_1(x), f_2(x), \ldots, f_p(x)) \\
\text{subject to } x \in X
\end{gathered}
\tag{8.36}
$$

When $p = 1$, this formulation represents a single objective optimization problem (SOOP); the problem is referred to as Bi-Objective Optimization Problems (BOOPs) if $p = 2$. The set X is given implicitly in the form of equality constraints or inequality constraints. Since most of the heuristic algorithms used to solve VRPs are not capable of solving MOOPs, the following five methods can transform them into SOOPs.

(1) The Weighted-Sum Approach

In the weighted-sum approach a weighted sum of the objective functions is minimized (Geoffrion 1968):

$$
\begin{gathered}
Min \sum_{k=1}^{p} w_k f_k(x) \\
\text{subject to } x \in X,
\end{gathered}
\tag{8.37}
$$

where w is the weight vector of p dimensions, which can also be considered as the penalty coefficient. The smaller the coefficient, the more important the objective is.

(2) The Weighted t-th Power Approach

In the weighted t-th power approach a weighted sum of the objective functions taken to the power of t is minimized (White 1988):

$$
\begin{gathered}
Min \sum_{k=1}^{p} w_k (f_k(x))^t \\
\text{subject to } x \in X,
\end{gathered}
\tag{8.38}
$$

where w is the weight vector of p dimensions and $t > 0$.

(3) The Weighted *t*-th Power Approach

Assuming that x^0 is a feasible solution of MOOP, based on the weighted-sum approach, the constraints related to all targets and the values of corresponding feasible solution x^0 are added to the model (Guddat et al. 1985):

$$\begin{aligned} & Min \sum_{k=1}^{p} w_k f_k(x) \\ & \text{subject to } f_k(x) \le f_k\left(x_k^0\right), k = 1, 2, \ldots, p \\ & \quad x \in X. \end{aligned} \tag{8.39}$$

(4) The ε-Constraint Approach

The ε-constraint approach only preserves one objective function while all the other objective functions generate new constraints. The *k*-th ε-constraint problem is formulated as (Chankong et al. 1985):

$$\begin{aligned} & Min \sum_{k=1}^{p} w_k f_k(x) \\ & \text{subject to } f_i(x) \le \varepsilon_i, i = 1, 2, \ldots, p, i \ne k \\ & \quad x \in X. \end{aligned} \tag{8.40}$$

(5) The Improved ε-Constraint Approach

The ε-constraint approach has numerical disadvantages when applied to problems with a specific structure, in particular discrete multi-objective problems. The improved ε-constraint method introduces a relaxation vector *l* to overcome those difficulties (Ehrgott and Ruzika 2008)

$$\begin{aligned} & Min f_k(x) - \sum_{i \ne k} w_i l_i \\ & \text{subject to } f_i(x) + l_i \le \varepsilon_i, i = 1, 2, \ldots, p; i \ne k \\ & \quad l_i \ge 0, i = 1, 2, \ldots, p; i \ne k \\ & \quad x \in X, \end{aligned} \tag{8.41}$$

and

$$
\begin{aligned}
& Minf_k(x) - \sum_{i \neq k} w_i l_i \\
& \text{subject to } f_i(x) - l_i \leq \varepsilon_i, i = 1, 2, \ldots, p; i \neq k \\
& \quad l_i \geq 0, i = 1, 2, \ldots, p; i \neq k \\
& \quad x \in X.
\end{aligned}
\tag{8.42}
$$

8.5 Discussions

There is no a single criterion regarding which method is better than the others because it depends considerably upon how we measure the risk of solutions. Circumstances around the application scenario, such as the level of road, traffic flows, and economic-social development state, also play an important role in risk assessment.

In fact, the risk of being rubbed is the most typical risk factor that the routing and scheduling of CIT vehicles differentiate from others. Robbers would consider a robbery successful if the financial gains meet or exceed their expectations if they escape or feel comfortable that they have not left any evidence that can lead to their identification and prosecution. The success mostly depends upon their ability to take control of events at the scene, that is, to manage effectively all the risks that could (and do) occur in the robbery. Robbers of CIT vehicles paid more attention to these than robbers of other types of premises. There were essentially four elements that made them good risk managers: they had experience which they were able to put to well use; they were motivated in a way that was more likely to avoid the dangers of not being in full control; they were much better prepared for their robberies; and they were willing to threaten and use violence (Gill 2001). These are types of crime facilitators' in that their existence enables the crime to occur. If these and other facilitators did not exist, then the crime may not occur or would be made more difficult. Therefore, in order to reduce the risk of the CIT vehicles being robbed, it is far from enough to do risk assessment on the routes. CIT sectors should take various measures to decrease the probability of robbery and the loss after a robbery. For example, assigning more armored staff for each vehicle, reinforcement of CIT vehicles, regular safety training for employees and so on.

8.6 Conclusions

In this study, we have introduced the origin and development of CTVRP models. Then, we reviewed the methods and models used for assessing risks in cash transportation. Finally, we designed a mathematical model for the CTVRP with uniform time windows and proposed the solution algorithms.

There are three categories of methods that have been used in the risk assessment in the cash-in-transit industry. In addition, we present a single-objective CTVRP model with minimizing the total operational cost and considering risk constraints for the transportation of valuable goods.

In order to solve the CTVRP model proposed in this study, many existing single-objective heuristic algorithms for VRPs can be used directly. As for multi-objective CTVRPs, heuristic algorithms similar to NSGA-II could be employed. Moreover, the single-objective heuristic algorithms also can be utilized before the objective functions are transformed into a single objective function.

Acknowledgements This research was funded by the Chongqing Graduate Scientific Research Innovation Project [grant number CYB20178].

References

Akgün V, Erkut E, Batta R (2000) On finding dissimilar paths. Eur J Oper Res 121(2):232–246

Androutsopoulos KN, Zografos KG (2012) A bi-objective time-dependent vehicle routing and scheduling problem for hazardous materials distribution. EURO J Transp Logist 1(1):157–183

Bell JE, McMullen PR (2004) Ant colony optimization techniques for the vehicle routing problem. Adv Eng Inform 18(1):41–48

Bozkaya B, Salman FS, Telciler K (2017) An adaptive and diversified vehicle routing approach to reducing the security risk of cash-in-transit operations. Networks 69(3):256–269

Bula GA, Murat Afsar H, González FA, Prodhon C, Velasco N (2019) Bi-objective vehicle routing problem for hazardous materials transportation. J Clean Prod 206:976–986

Chankong V, Haimes Y, Thadathil J, Zionts S (1985) Multiple criteria optimization; a state of the art review. Decis Mak Mult Object 36–90

Chen Y, Cowling P, Polack F, Remde S, Mourdjis P (2017) Dynamic optimisation of preventative and corrective maintenance schedules for a large scale urban drainage system. Eur J Oper Res 257(2):494–510

Dantzig GB, Ramser JH (1959) The truck dispatching problem. Manage Sci 6(1):80–91

Ehrgott M, Ruzika S (2008) Improved e-constraint method for multiobjective programming. J Optim Theory Appl 138(3):375

Geoffrion AM (1968) Proper efficiency and the theory of vector maximization. J Math Anal Appl 22(3):618–630

Geyer HS (2007) International handbook of urban policy: contentious global issues. Edward Elgar Publishing

Ghaderi A, Burdett RL (2019) An integrated location and routing approach for transporting hazardous materials in a bi-modal transportation network. Transp Res Part E: Logist Transp Rev 127:49–65

Ghannadpour SF, Zandiyeh F (2020) A new game-theoretical multi-objective evolutionary approach for cash-in-transit vehicle routing problem with time windows (A real life case). Appl Soft Comput 93:106378

Gill M (2001) The craft of robbers of cash-in-transit vans: crime facilitators and the entrepreneurial approach. Int J Sociol Law 29(3):277–291

Gillett BE, Miller LR (1974) A Heuristic algorithm for the vehicle-dispatch problem. Oper Res 22(2):340–349

Guddat J, Guerra Vasquez F, Tammer K, Wendler K (1985) Multiobjective and stochastic optimization based on parametric optimization. Math Res 26

Guerrero WJ, Sarmiento-Lepesqueur A, Martínez-Agaton C (2020) Cash distribution model with safety constraints. In: International conference on computational logistics. Springer

Hepenstal S, Johnson SD (2010) The concentration of cash-in-transit robbery. Crime Prev Community Saf 12(4):263–282

Hoogeboom M, Dullaert W (2019) Vehicle routing with arrival time diversification. Eur J Oper Res 275(1):93–107

IBRD (2018) An overview of financial inclusion. https://www.worldbank.org/en/topic/financialinclusion/overview#1 Last Accessed 8 Mar 2021

Jemai J, Zekri M, Mellouli K (2012) An NSGA-II algorithm for the green vehicle routing problem. In: European conference on evolutionary computation in combinatorial optimization. Springer

Karakatic S, Podgorelec V (2015) A survey of genetic algorithms for solving multi depot vehicle routing problem. Appl Soft Comput 27:519–532

Kazantzi V, Kazantzis N, Gerogiannis VC (2011) Risk informed optimization of a hazardous material multi-periodic transportation model. J Loss Prev Process Ind 24(6):767–773

Kumar A, Roy D, Verter V, Sharma D (2018) Integrated fleet mix and routing decision for hazmat transportation: a developing country perspective. Eur J Oper Res 264(1):225–238

Letchford AN, Lysgaard J, Eglese RW (2007) A branch-and-cut algorithm for the capacitated open vehicle routing problem. J Oper Res Soc 58(12):1642–1651

Lin N, Shi Y, Zhang T, Wang X (2019) An effective order-aware hybrid genetic algorithm for capacitated vehicle routing problems in internet of things. IEEE Access 7:86102–86114

Martí R, Luis González Velarde J, Duarte A (2009) Heuristics for the bi-objective path dissimilarity problem. Comput Oper Res 36(11):2905–2912

Ngueveu SU, Prins C, Calvo RW (2009) A hybrid tabu search for the m-peripatetic vehicle routing problem. Springer, Matheuristics, pp 253–266

Parsafard M, Esmaeel A, Masoud K, Mohammadreza N, Li X (2015) Practical approach for finding optimum routes for fuel delivery trucks in large cities. Transp Res Rec 2478(1):66–74

Pradhananga R, Taniguchi E, Yamada T (2010) Ant colony system based routing and scheduling for hazardous material transportation. Procedia Soc Behav Sci 2(3):6097–6108

Pradhananga R, Taniguchi E, Yamada T, Qureshi AG (2014a) Bi-objective decision support system for routing and scheduling of hazardous materials. Socioecon Plann Sci 48(2):135–148

Pradhananga R, Taniguchi E, Yamada T, Qureshi AG (2014b) Environmental analysis of pareto optimal routes in Hazardous material transportation. Procedia Soc Behav Sci 125:506–517

Radojicic N, Djenic A, Maric M (2018a) Fuzzy GRASP with path relinking for the risk-constrained cash-in-transit vehicle routing problem. Appl Soft Comput 72:486–497

Radojicic N, Maric M (2018b) Some possibilities of using fuzzy logic within methods for solving a variant of vehicle routing problem. In: 2018 IEEE 18th international symposium on computational intelligence and informatics (CINTI). IEEE

Radojicic N, Maric M, Takaci A (2018) A new fuzzy version of the risk-constrained cash-in-transit vehicle routing problem. Inf Technol Control 47(2):321–337

Russo F, Rindone C (2011) Planning in road evacuation: classification of exogenous activities. WIT Trans Built Environ 116:639–651

Santos FA, Mateus GR, da Cunha AS (2015) A branch-and-cut-and-price algorithm for the two-echelon capacitated vehicle routing problem. Transp Sci 49(2):355–368

Soriano A, Vidal T, Gansterer M, Doerner K (2020) The vehicle routing problem with arrival time diversification on a multigraph. Eur J Oper Res 286(2):564–575

Stanojevic M, Stanojevic B, Vujoševic M (2013) Enhanced savings calculation and its applications for solving capacitated vehicle routing problem. Appl Math Comput 219(20):10302–10312

Talarico L, Sörensen K, Springael J (2015a) The k-dissimilar vehicle routing problem. Eur J Oper Res 244(1):129–140

Talarico L, Sörensen K, Springael J (2015b) Metaheuristics for the risk-constrained cash-in-transit vehicle routing problem. Eur J Oper Res 244(2):457–470

Talarico L, Sörensen K, Springael J (2017) A biobjective decision model to increase security and reduce travel costs in the cash-in-transit sector. Eur J Oper Res 24(1–2):59–76

Talarico L, Springael J, Sörensen K, Talarico F (2017) A large neighbourhood metaheuristic for the risk-constrained cash-in-transit vehicle routing problem. Comput Oper Res 78:547–556

Tarantilis CD, Kiranoudis CT (2004) An adaptive memory programming method for risk logistics operations. Int J Syst Sci 35(10):579–590

Tikani H, Setak M, Demir E (2020) A risk-constrained time-dependent cash-in-transit routing problem in multigraph under uncertainty. Eur J Oper Res

Tikani H, Setak M, Demir E (2021) Multi-objective periodic cash transportation problem with path dissimilarity and arrival time variation. Expert Syst Appl 164:114015

Vanhove S (2012) Alternative routing algorithms for road networks. Ghent University

White DJ (1988) Weighting factor extensions for finite multiple objective vector minimization problems. Eur J Oper Res 36(2):256–265

Xu G, Li Y, Szeto WY, Li J (2019) A cash transportation vehicle routing problem with combinations of different cash denominations. Int Trans Oper Res 26(6):2179–2198

Yan S, Wang S-S, Chang Y-H (2014) Cash transportation vehicle routing and scheduling under stochastic travel times. Eng Optim 46(3):289–307

Yan S, Wang S-S, Wu M-W (2012) A model with a solution algorithm for the cash transportation vehicle routing and scheduling problem. Comput Ind Eng 63(2):464–473

Chapter 9
Mitigating Energy Supply Risks: Factors Identification and Pathway Selection for China's Renewable Energy Development Fuzzy PROMETHEE

Linmao Ma and Long Zhang

Abstract Renewable energy plays a vital role in promotion of sustainable transition and enhancement of energy security, while various factors have brought risks that threaten renewable energy development. Great attentions have been paid on investigating the barriers to the innovation and application of novel energy technologies, while the risks and inhibitors resulting in its social acceptance have been ignored. This study reviewed the existing literature on the factors and risks regarding renewable energy development and summarized them from the four aspects including environment, economy, social politics, and technology, with the aim of constructing a general framework for multi-criteria decision analysis on the evaluation of renewable energy development. Based on that, the potential of various renewable energy technologies for mitigating energy supply risks in China have been evaluated. To facilitate that, the approach of Fuzzy PROMETHEE integrating the Fuzzy Set theory with the PROMETHEE method was adapted to deal with this real-world problem in this study, and fuzzy linguistic variables were used to assist experts in making their judgments on the relative performance of various renewable energy technologies to avoid producing imprecise information. Finally, the results showed that wind energy, solar photovoltaic, and biomass are the preferable energy sources presently for mitigating energy supply risks according to the current status of each renewable energy technology.

Keywords Energy security risks · Renewable energy technology · Multi-criteria decision analysis · Fuzzy PROMETHEE

L. Ma
School of Management, South-Central University for Nationalities, Wuhan 430074, China

L. Zhang (✉)
School of Business, Xinyang Normal University, No. 237 Nanhu Road, Xinyang 464000, China
e-mail: longzh@xynu.edu.cn

J. Ren (ed.), *Multi-Criteria Decision Analysis for Risk Assessment and Management*,
Industrial Ecology and Environmental Management 1,
https://doi.org/10.1007/978-3-030-78152-1_9

9.1 Introduction

In recent decades, there has been an increasing interest in renewable energy which is often associated with mitigation energy supply risks, on of clean electricity and no significantly harmful impacts on environment. A plenty of researches on renewable energy have been undertaken, and most of them drew the conclusion that renewable energy can play a vital role in future energy systems for the improvement of energy security from the perspectives of environment protection and economic development because of its environmental benefits compared with fossil energy sources (He 2016; Rahman 2017; Shaaban and Petinrin 2014; Zhang 2017, 2021). The booming Chinese economy pushes the rapid growth of energy consumption in China. In 2010, it surpassed the US, becoming the world's largest energy consumer, with a share of 24 percent of world total energy consumption and 77% of the global net increase in energy consumption in 2019 (Bob 2015). The increasing demand for energy has triggered the public scrutiny on the sustainability issues associated with the risks on energy supply. Although coal has always been China's superior resource, it is becoming increasingly relying on energy imports to sustain energy supply, which increases the risks of unexpected energy supply disruptions, and China's dependence on foreign oil has been even over 80% in 2019. At the same time, about 28% of the world total carbon dioxide emissions were produced in China, which also bring risks to environmental sustainability. However, the global climate change requires less energy consumption, especially fossil fuels. Furthermore, the consumption of fossil fuels causes high loading of atmospheric pollutants, leading to common urban haze in major cities in eastern China, which has greatly threatened public health (Zeng 2015). In order to reduce the environmental risks associated with the energy sector, renewable energy, such as solar energy, biomass, wind power, and hydropower, has been prompted to mitigate environment pollution.

Renewable energy development has been influenced by a number of factors, including economic, technological, political and legal, social, and environmental aspects. While, most researchers have been focusing on the technical and economic aspects of renewable energy but ignored the social aspects (Campbell 2018; Ozonoh 2018). According to the statistic from Web of Science, there have been 57,338 papers related to renewable energy by 2020 with 197 papers involving social science and social issue. It is worth noting that the top three research subjects include energy fuels, engineering, and science technology, which indicates that most scholars focus on the technological aspect of renewable energy development while ignoring the other aspects especially the social and political factors. Although the research and development of renewable energy technologies is important for promoting renewable energy development, the risks from other factors, for instance, the economic feasibility, whether or not the public is willing to accept it, also determines the success of renewable energy technologies. Considered that technological, economic, and environmental factors have been investigated in a lot of studies, further researches should be conducted to include the influence of risky factors in social aspect in renewable energy development.

Fortunately, some scholars have shed light on the social factors affecting renewable energy development, especially on their possible risks in social acceptance and public awareness. Wüstenhagen et al. developed a three-dimension model for the social acceptance of renewable energy, i.e., socio-political acceptance, community acceptance, and market acceptance which covered the social, political, environmental, and economic aspects of social acceptance (Wüstenhagen et al. 2007). Apparently, it is hard to achieve the goal of renewable energy development without the active participation and acceptance of the public. Therefore, social acceptance should be taken into consideration during the promotion of renewable energy technologies. The social acceptance to renewable energy technologies among residents is influential not only to the renewable energy project itself but also to the success of sustainable development of that region. Furthermore, Conrad and Busch directly highlighted that renewable energy is not only a technical issue but also one for the social sciences (Ribeiro et al. 2011). Thus, there is an increasing level of attention paid to the social acceptance of renewable energy. Upham et al. conducted a large-scale study in the United Kingdom to investigate the main theoretical issues in relation to social acceptance Upham et al. (2009). Ribeiro et al. also stressed the inclusion of the social dimension in the planning of power systems as an issue (Ribeiro et al. 2011). In fact, the social acceptance to renewable energy projects is not always clear and certain. The local residents usually reject these projects close to their community, which is frequently described as "a not in my backyard (NIMBY)" syndrome, and would lead to an attitude-behavior gap during the implementation of the wind power (Devine-Wright 2013). The strong public support for wind energy expressed in opinion surveys did not lead to the corresponding success rate achieved in planning applications for wind power developments (Bell et al. 2005).

As to renewable energy development in China, the largest proportion of the renewable energy is contributed by hydropower with a share of 16.9% in the total generated energy and followed by wind power. With improved technology and clear guiding policies, Chinese wind power manufacturers are growing fast, and wind power would have a promising future in China in terms of available resources and markets (Feng 2015). Unfortunately, parts of the potential energy generated by the wind turbines could not be transmitted to the electric power grid in China, leading to a serious wind energy rejection problem (Zhang 2016). For the solar technology, a quantitative approach was adopted to investigate the social acceptance to solar energy technologies from end users' perspective, which shown that there was a considerable high level of social acceptance and public awareness of the solar water heater while another major application of solar energy, solar PV, did not gain a high level of social acceptance or public awareness (Yuan et al. 2011). Additionally, the development of other renewable energy sources also faces a great challenge. For instance, the first national offshore wind power project is connected to the grid in 2010 (Zhang 2016); however, Li et al. pointed out that China is still lacking experience in this industry from the aspect of technology (Li 2014). The application of waste-to-energy treatment from municipal solid waste also faced strong protest by local communities, especially in cities with high population densities due to the public protest to accept (Ren and Sovacool 2015). Hu et al. analyzed what deterred the development of renewable energy sources including solar, wind, geothermal, and biomass in Southwest China from the view of geographical position, land resources, and technology (Hu 2016). Obviously, it is not adequate to just investigate the factors influencing renewable energy development from the view of a specific aspect because the interrelationship among these factors is very complex, involving the technology elements, economic limitation, social acceptance, and public awareness and attitudes.

Due to the importance of social aspects in promoting renewable energy development, different studies have shown that the socio-economic characteristics of the population clearly influence public acceptance to renewable energy technologies, and they may bring additional risks to renewable energy development. Aaen et al. provided a deep insight into the process of the citizens' sense making, which could improve and overcome shortcomings in the current thinking about public participation (Aaen et al. 2016). By comparing the implementation of both large-scale and small-scale projects, Capodaglio et al. concluded that the introduction of small-scale projects would not confront problems regarding public acceptance as severe as large-scale projects (Capodaglio et al. 2016). Specifically, from the aspect of personal traits, education and income play important roles in public awareness of renewable energy, which shown that efforts to educate the attentive public are needed for the improvement of public participation in the development of wind and solar power (Yuan et al. 2011; Echegaray 2014). Bertsch et al. also stated that age and education were the most relevant factors determining public participation in renewable energy development (Bertsch 2016). Further studies demonstrated that the level of energy knowledge and literacy are related to public acceptance to renewable energy, while some other researches have shown that people perceive technologies as contributing to lower prices (Sovacool and Blyth 2015; Ribeiro 2014). From the perspective of policy, Wolsink pointed out that it helps to improve the social acceptance to environmental policies by engaging local residents into the policy making process (Wolsink 2010). Public participation and the choice of residents for renewable energy play an important role in the successful adoption of wind energy technology, and the perception of political processes and distributive justice would influence the public preferences related to renewable energy at the same significant level (Langer 2016).

What is more, different kinds of renewable energy technologies confront distinct risks resulting from the social aspect. Currently, the residents appear to be more willing to accept wind energy as shown in the following literature. Based on the evaluation of the level of knowledge on wind and solar power and the investigation on the social acceptance to renewable energy, the result demonstrated that there is a high level of acceptance to solar and wind power in Brazil (Sena et al. 2016). Rensburg et al. explored what influences the approval of wind farm planning and found that proximity to dwellings, towns, or protected habitats does not influence planning outcomes (Rensburg et al. 2015). Similarly, Brennan and Rensburg revealed that most local residents were willing to make tradeoffs to allow for wind power if provision is made for a community representative and setback distance is increased (Brennan and Rensburg 2016). In addition to the onshore wind farm, some scholars also focused on the offshore wind farm. Firestone et al. studied how public acceptance of offshore wind power change from 2005 to 2009 and found that the desire for energy independence is an increasingly significant reason for supporters and has motivated some individuals to switch from opposition to support (Firestone 2012). And then, Kermagoret et al. highlighted that developing offshore wind farms can enhance territorial integrity and its heritage dimension so as to improve social acceptance (Kermagoret 2016).

Based on the existing literature, it is obvious that risks related to social factors renewable energy are very important, but there are no good enough investigations on them in China. What is more, current studies are limited to simple identification and description of the potential risk factors influencing social acceptance to renewable energy, and a few scholars have focused on the causality between these factors and how these factors affect public acceptance. However, this is a complex and involved system comprising various personal and social issues such as age, education, gender, income, energy knowledge and literacy, and the level of political participation. Description of such a system requires a proper approach to deal with the difficulty arising from the dynamic nature of these influences among the factors. Furthermore, some scholars have demonstrated that social factors are crucial for promoting renewable energy development. Nevertheless, it is confused to depict the processes while all these social factors are taken into account during the political decision-making process of renewable energy development. Therefore, it is necessary to develop a framework and methods involving the social aspect and public awareness for the evaluation of key factors affecting the potential risks for renewable energy development and conduct an empirical study in China in order to mitigate renewable energy supply risks. To enhance the energy security and reduce energy supply risks through the promotion of renewable energy, a shared understanding of the potential of various renewable energy technologies has to be built to ultimately result in more reasonable assessments for renewable energy development. Eventually, this research intends to use the fuzzy preference ranking organization method for enrichment evaluation (Fuzzy-PROMETHEE) to solve these two problems.

The rest of this paper is organized as follows: Sect. 9.2 analyzes the current research on the elements affecting renewable energy development and constructs an analysis framework for the investigation on various renewable energy options' potential for coping with energy supply risks. Section 9.3 introduces the basic information about the method of PROMETHEE as well as the fuzzy PROMETHEE. In Sect. 9.4, a case study on the development of Chinese renewable energy is conducted using the aforementioned method. Section 9.5 concludes this paper through the analysis of the results.

9.2 Potential Risks Affecting Renewable Energy Development

With the unremitting and continuing efforts of both governments and enterprises, remarkable progress has been achieved in renewable energy development around the world. However, there are still many obstacles in the implementation of renewable energy projects, which have increased the risks in renewable energy development. Researchers have conducted plenty of investigations on the risks and elements affecting renewable energy development from different perspectives in their own domains, including technological, economic, political, and environmental factors. This section summarized the existing studies on this issue to exhibit the complexity of these elements and built an integrated analysis framework to further the evaluation of the renewable energy potential thoroughly.

9.2.1 Technological Risks

The elements involved in the technological dimension still pose great risks for the failure of renewable energy development and consists of the lack of proper production device adapted to a specific environment, the risk of operation and maintenance, the risk of equipment stability and economy, and the absence of technological operators. The technological risks or factors have been considered as the key for wind power development in some countries (Richards et al. 2012). Hu et al. drew the same conclusion that the current technology for wind power generation cannot adapt to the climate in the specific region and concurrently emphasized that biomass energy conversion technology has a great influence on its utilization through a comprehensive investigation on the development of solar energy, wind power and biomass considering geographical location, land resources, and technical factors (Hu 2016). Jacobson and Johnson focused on the impact of R & D of solar energy-related technologies on the growth of the solar energy industry and pointed out that innovation on the relevant technology is necessary for its promotion (Jacobsson and Johnson 2000). Similarly, the R & D of other renewable energy technologies also need to be enhanced (Zhang 2013). Apart from technological innovation, the lack of relevant experience in project implementation may bring risks that lead to the failure of renewable energy projects (Li 2014).

9.2.2 Economic Risks

In addition to the limitation of technological elements, the promotion of renewable energy is also restricted by economic risks. First of all, the high capital investment and construction cost have led to great risks which seriously prevent energy enterprises from investing in renewable energy projects (Ackermann and Söder 2002; Painuly 2001; Eleftheriadis and Anagnostopoulou 2015). Therefore, the government's financial subsidy system and corresponding fiscal stimulus are needed for promoting its development (Eleftheriadis and Anagnostopoulou 2015). Secondly, the local infrastructure also has impacts on renewable energy development, the most important of which is the transportation of raw materials and equipment (Hu 2016), and the impact of existing power grid construction on renewable energy power generation also needs to be considered (Zhang 2016). Finally, the absence of economic benefits from renewable energy production is also taken as a major risk to renewable energy development (Nadaï and Labussière 2009; Zhang et al. 2012). Besides, the construction and regulation of the renewable energy market are also related to economic risks, and renewable energy consumption should be stimulated and promoted to enhance the market competitiveness of renewable energy (Zhang 2013; Roques et al. 2010, 2012).

9.2.3 Social-Political Risks

The political atmosphere regarding energy policy and institutions is also a potential risk in renewable energy development, and the proper public policy intervention plan is a key element (Jagadeesh 2000). The research of Wolsink showed that institutional factors have more influence on renewable energy development than others (Wolsink 2000). Richards also found that the factor with respect to the renewable energy policy is the most important one for the adoption of renewable energy except the technical elements (Richards et al. 2012). The potential impact of social-political factors on renewable energy development include the establishment of the regulatory agency, the legal and regulatory system for renewable energy development, professional institutions aiming at conducting researches on policy effect, and the process of renewable energy policy-making considering the participation of other stakeholders (Painuly 2001). Breukers and Wolsink have studied the impact of institutional factors on wind energy development, and the results indicated that the establishment of an institutionalized policy to ensure residents' participation in project planning can effectively alleviate the residents' negative attitude toward these projects (Breukers and Wolsink 2007). In the study of China's solar photovoltaic market, it is also found that the development of the solar photovoltaic market seriously depends on the central policy so that it is necessary to coordinate the policy objectives and incentive measures between the central and local governments (Corwin and Johnson 2019). In addition to the related policies and regulations issued by the government, the lack of public access to reliable information on renewable energy products and technologies also hinders the growth of the renewable energy market (Painuly 2001; Reddy and Painuly 2004; Luthra 2015). It is quite necessary for the government to release information about the necessity of developing renewable energy and the performance of renewable energy products and technologies to the residents in order to mitigate the related risks and promote the adoption of renewable energy (Agterbosch et al. 2007). Sovacool once again proved that the lack of information about renewable energy and the existing energy policies hinder the utilization of novel technology (Sovacool 2009).

9.2.4 Environmental Risks

The potential risks of renewable energy production on the local environment are also vital factors obstructing renewable energy development. The production of renewable energy inevitably yields some by-products with an adverse risk on residents' health and the local environment from the operation of its devices, which directly affects the implementation of renewable energy projects. Hernandez et al. studied the risks of solar photovoltaic on local biodiversity, water resources, public health, air pollution, soil degradation, ecological landscape, land use, and vegetation change, and those side-effects on the local environment seriously affect renewable energy development (Hernandez 2014). Hu et al. also emphasized that land occupation in renewable energy development planning would conflict with traditionally cultivated land, which would also damage the residents' awareness toward renewable energy (Hu 2016); Warren et al. investigated various factors affecting wind energy development, which showed that the construction of wind power plants encountered many difficulties due to the serious impact on local landscape, animals, plants, and noise pollution (Warren 2005). Similarly, renewable energy development may lead to the reduction of forest area, and bring about impacts including noise, light, visual, and landscape on residents, which are viewed as environmental issues and conversely have a passive influence on its growth (Rensburg et al. 2015; Devine-Wright 2005, 2010; Mann and Teilmann 2013; Wolsink 2007). In addition to the environmental risks caused by the production of renewable energy, scholars also began to focus on individual social factors, suggesting that public understanding of renewable energy and their own environmental awareness also have latent risks on the adoption and implementation of local renewable energy projects (Mann and Teilmann 2013; Wolsink 2007; Saidur 2011; Bright 2008).

9.2.5 A General Framework of Risks Affecting Renewable Energy Development

In summary, based on the review of the existing literature on risks and factors related to renewable energy development, it can be found that most studies are conducted from a distinct perspective with each other, although all of them have focused on the factors influencing renewable energy development. Therefore, this paper reviewed these researches in detail and sorted out the corresponding influencing factors according to the following dimensions: environmental factors, social and political factors, economic factors, and technical factors, as shown in Table 9.1. This classification includes the technical, economic, social-political, social and cultural, information, and environmental risks in the existing literature. It is also convenient to further study these risks and evaluate the potential of each renewable energy source.

Table 9.1 Various risks from four dimensions affecting renewable energy development

Dimension	Indicator	References
Environmental	En1: Impact on the local animal and plants	Devine-Wright (2005), Hernandez et al. (2014), Saidur et al. (2011), Magoha (2002), Bright et al. (2008), Chiras et al. (2009)
	En2: Noise impacts	Manwell et al. (2010), Mann and Teilmann (2013), Wolsink (2007), Mercer et al. (2017)
	En3: Visual externality and landscape impacts	Rensburg et al. (2015), Wolsink (2007), Hernandez et al. (2014)
	En4: Environment awareness	Devine-Wright (2007), Demski et al. (2014), Greenberg (2009), Hobman and Ashworth (2013), Ertör-Akyazı et al. (2012)
	En5: Energy literature	Komarek et al. (2011), Mallett (2007), Aitken (2010), Ladenburg and Möller (2011), Cicia et al. (2012), Mercer et al. (2017)
	En6: Degree of occupation of local resources (land, water, forest resources)	Sena et al. (2016)
	En7: Location of the project	Nadaï and Labussière (2009), Rensburg et al. (2015)
Social-political	SP1: Experience on the development of renewable energy	Nadaï and Lasière (2009), Soland et al. (2013), Richard et al. (2012)
	SP2: Number of existing renewable energy project	Jobert et al. (2007), Richard et al. (2012), Wizelius (2015)
	SP3: Policy and regulation	Wolsink (2000), Çoban and Onar (2017), Breukers and Wolsink (2007)
	SP4: Degree of public participation	Painuly (2001), Zhang et al. (2012)
	SP5: Access to relevant information	Painuly (2001), Reddy and Painuly (2004), Luthra et al. (2015), Asadullah (2014)
Economic	Ec1: Impact on the local real estate	Jobert et al. (2007)
	Ec2: Investment cost	Ghimire and Kim (2018), Painuly (2001), Eleftheriadis (2015), Zhang et al. (2012), Zhang et al. (2013)
	Ec3: Market acceptance	Roques and Hiroux (2010), Sovacool and Ratan (2012), Richard et al. (2012), Luthra et al. (2015)
	Ec4: Energy price	Sovacool and Ratan (2012), Frantál (2015), Çoban and Onar (2017)

(continued)

Table 9.1 (continued)

Dimension	Indicator	References
	Ec5: Supply chain and transportation	Hu et al. (2016), Painuly (2001), Ghimire and Kim (2018), Mittal et al. (2018)
	Ec6: Community welfare	Nadaï and Labussière (2009), Zhang et al. (2012), Liu and Pistorius (2012)
Technological	Te1: Mature of technology	Yuan et al. (2011), Li et al. (2014), Çoban and Onar (2017), Richard et al. (2012), Mercer et al. (2017), Mittal et al. (2018)
	Te2: Availability of technology	Ghimire and Kim (2018), Hu et al. (2016), Eleftheriadis (2015)
	Te3: Criterion of technology	Painuly (2001), Zhang et al. (2012), Asadullah (2014)
	Te4: The statue of operators	Ghimire and Kim (2018), Painuly (2001), Eleftheriadis, (2015), Luthra et al. 2015)
	Te5: Scale of energy project	Huang et al. (2013), Capodaglio et al. (2016)

9.3 Method

This section briefly presents the basic theory of fuzzy set and the method of PROMETHEE, as well as the hybrid model of fuzzy PROMETHEE.

9.3.1 *Fuzzy Set and Fuzzy Numbers*

The fuzzy set is proposed to deal with the vagueness of decision-making under an uncertain environment and is commonly used with linguistic variables. The fuzzy set theory was firstly introduced in 1965 (Zadeh 1965) and widely used in the solution of modern scientific decision-making and fuzzy optimal problems with vague and imprecise variables. According to the definition of membership function, the fuzzy number has various forms, such as the triangular fuzzy number and trapezoidal fuzzy number. Suppose A is a trapezoidal fuzzy number on the universe the membership function is shown as Eq. (9.1):

$$\mu_A(x) = \begin{cases} 0, & x < l \\ (x-l)/(m-l), & l \le x \le m \\ 1, & m \le x \le n \\ (r-x)/(r-n), & n \le x \le r \\ 0, & x > r \end{cases} \tag{9.1}$$

Table 9.2 The algebraic operation with fuzzy numbers (Goumas and Lygerou 2000; Gul et al. 2018)

Operation	Formulation
Addition	$A_1 \oplus A_2 = [l_1, m_1, n_1, r_1] \oplus [l_1, m_2, n_2, r_2]$ $= [l_1 + l_2, m_1 + m_2, n_1 + n_2, r_1 + r_2]$
Opposite	$-A_1 = [-l_1, -m_1, -n_1, -r_1]$
Subtraction	$A_1 \ominus A_2 = [l_1, m_1, n_1, r_1] \ominus [l_1, m_2, n_2, r_2]$ $= [l_1 - r_2, m_1 - n_2, n_1 - m_2, r_1 - l_2]$
Multiplication	$A_1 \otimes A_2 = [l_1, m_1, n_1, r_1] \otimes [l_1, m_2, n_2, r_2]$ $= [l_1 \times l_2, m_1 \times m_2, n_1 \times n_2, r_1 \times r_2]$
Division	$A_1 \oslash A_2 = [l_1, m_1, n_1, r_1] \oslash [l_1, m_2, n_2, r_2]$ $= [l_1/l_2, m_1/m_2, n_1/n_2, r_1/r_2]$

where m and n are the most promising interval of the fuzzy number indicating the numbers belong with certainty to the set of available values, and l and r are the lower and upper limitation of the fuzzy number, respectively. This fuzzy number is denoted by the notion $A = [l, m, n, r]$. Correspondingly, the basic operations of two trapezoidal fuzzy number, $A_1 = [l_1, m_1, n_1, r_1]$ and $A_2 = [l_2, m_2, n_2, r_2]$, are list in Table 9.2 (Goumas and Lygerou 2000; Gul et al. 2018).

9.3.2 *The Method of PROMETHEE*

The method of PROMETHEE was proposed by Brans and Vincke in 1985, which were named as PROMETHEE-I (Brans and Vincle 1985), and then it was improved by Brans et al. to form a multi-criteria decision-making method of PROMETHEE-II (Brans et al. 1986). The basic principle of the PROMETHEE method is to select a more feasible solution according to the preference of different alternatives given by the decision-makers under multiple and complex criteria. The key to this method is to define a preference function for each attribute according to the preference of decision-makers and the requirements which are used to determine the preference indices between these alternatives. According to the outranking relation for all alternatives, the outflows and inflows are calculated to obtain the net flow for each alternative, which is used to determine the ranking of the alternative. The basic procedure of PRMETHEE is as follows (Goumas and Lygerou 2000; Gul et al. 2018):

Step 1: Determine the criteria set $C = \{c_1, c_2, \cdots, c_n\}$ for ranking the alternatives and the potential alterative set $A = \{a_1, a_2, \cdots, a_m\}$.

Step 2: Obtain a decision matrix from experts. This decision matrix is derived from the evaluation of all experts invited in the research on the alternatives according to their own preferences of and experiences toward these factors. The relative importance of the compared alternatives based pairwise comparison of alternatives is formulated as Eq. (9.2):

$$\widetilde{D} = \left[a_{ij}\right]_{m\times n} = \begin{bmatrix} a_{11} & a_{12} & \cdots & a_{1k} \\ a_{21} & a_{22} & \cdots & a_{2k} \\ \cdots & \cdots & \cdots & \cdots \\ a_{m1} & a_{m2} & \cdots & a_{mk} \end{bmatrix} i = 1, 2, \cdots, m; j = 1, 2, \cdots, n \tag{9.2}$$

Step 3: Select a preference function with respect to the feature of various criteria. There are six preference functions commonly used in the relevant researches, as shown in Eqs. (9.3)–(9.8).
Type 1: Usual criterion

$$P_k(d_k(a_i, a_j)) = \begin{cases} 0 \ d_k(a_i, a_j) < 0 \ (indifference) \\ 1 \ d_k(a_i, a_j) \geq 0 \ (strictindifference) \end{cases} \tag{9.3}$$

Type 2: Quasi-criterion

$$P_k(d_k(a_i, a_j)) = \begin{cases} 0 \ d_k(a_i, a_j) < p \ (indifference) \\ 1 \ d_k(a_i, a_j) \geq p \ (strictindifference) \end{cases} \tag{9.4}$$

Type 3: Criterion with linear preference

$$P_k(d_k(a_i, a_j)) = \begin{cases} d_k(a_i, a_j)/p \ 0 < d_k(a_i, a_j) < p \ (indifference) \\ 1 \qquad d_k(a_i, a_j) \geq p \ (strictindifference) \end{cases} \tag{9.5}$$

Type 4: Level criterion

$$P_k(d_k(a_i, a_j)) = \begin{cases} 0 \quad 0 < d_k(a_i, a_j) \leq q \ (indifference) \\ 0.5 \ q < d_k(a_i, a_j) <\leq p \\ 1 \quad d_k(a_i, a_j) \geq p \ (strictindifference) \end{cases} \tag{9.6}$$

Type 5: Criterion with linear preference and indifference area

$$P_k(d_k(a_i,a_j)) = \begin{cases} 0 & 0 < d_j(a_i,a_j) \le q \ (indifference) \\ \frac{d_k(a_i,a_j)-q}{p-q} & q < d_k(a_i,a_j) <\le p \\ 1 & d_k(a_i,a_j) \ge p \ (strictindifference) \end{cases} \tag{9.7}$$

Type 6: Gaussian criterion

$$P_k(d_k(a_i,a_j)) = \begin{cases} 0 & d_k(a_i,a_j) < 0 \ (indifference) \\ 1 - e^{-d_k(a_i,a_j)/2\delta^2} & d_k(a_i,a_j) \ge 0 \ (strictindifference) \end{cases} \tag{9.8}$$

Step 4: Compute the preference index for each alterative to distinct its performance. The preference index is computed by Eq. (9.9):

$$\pi(a_i,a_j) = \sum_{j=1}^{n} P_k(a_i,a_j), \text{ where}$$

$$a_i, a_j \in A \quad k = 1, 2, \cdots, n \tag{9.9}$$

Step 5: Generate the entering and leaving flow for each alterative by Eqs. (9.10)–(9.11).

$$\text{Leaving flow: } \phi^+(a_i) = \frac{1}{m-1} \sum_{a_j \ne a_i}^{m} \pi(a_i,a_j) \quad a_i, a_j \in A \tag{9.10}$$

$$\text{Entering flow: } \phi^-(a_i) = \frac{1}{m-1} \sum_{a_j \ne a_i}^{m} \pi(a_j,a_i) \quad a_i, a_j \in A \tag{9.11}$$

The leaving flow $\phi^+(a_i)$ denotes the preference of the alternative a_i over all the other alternatives, so the leaving flow with a higher value means that the alternative is relatively more important than other alternatives. By contrast, the value of entering flow $\phi^-(a_i)$ indicates the preference of all other alternatives compared to the alternative a_i, and its shows the relative importance of other alternatives. The alternative with a higher entering flow value is inferior.

Step 6: Rank the alternatives according to the rule of PROMETHEE-I/II.

In the method of PROMETHEE-I, alternative a_i strongly outranks alternative a_j when the leaving flow for a_i is greater than the value of a_j and the entering flow for a_i is less than the entering flow for a_j; alternative a_i is indifferent to alternative a_j under the situation that the leaving and entering flow for a_i is equivalent to the corresponding value of alternative a_j; otherwise, alternatives a_i and a_j are treated as incomparable. As indicated by Eqs. (9.12) and (9.13).

$$a_i \ outranking \ a_j \ iff \ \phi^+(a_i) \geq \phi^+(a_j) and \phi^-(a_i) \leq \phi^-(a_j) \tag{9.12}$$

$$a_i \ is \ indifferent \ a_j \quad iff \ \phi^+(a_i) = \phi^+(a_j) and \phi^-(a_i) = \phi^-(a_j) \tag{9.13}$$

As to the method of PROMETHEE-II, the net flow $\phi(a_i) = \phi^+(a_i) - \phi^-(a_i)$ is calculated to rank the alternatives providing a complete order of alternatives compare with the partial order produced by PROMETHEE-I. The alternatives with higher net flow outperform other alternatives with lower value, as presented by Eqs. (9.14) and (9.15).

$$a_i \ outranking \ a_j \quad iff \ \phi(a_i) = \phi(a_j) \tag{9.14}$$

$$a_i \ is \ indifferent \ a_j \quad iff \ \phi(a_i) = \phi(a_j) \tag{9.15}$$

9.3.3 Procedure of Fuzzy PROMETHEE

Generally, the method of PROMETHEE is suitable for assisting the decision-makers in solving complex decision problems in a multi-criteria situation. However, for the process of decision-making, the decision makers are usually required to give a crisp value for the specific practical problem with their own knowledge and experience on this issue, but encountering a fuzzy and complex problem makes it difficult for the decision-makers to do so. In this situation, the decision-makers in the analysis can only provide a vague evaluation for all alternatives to a certain degree in a specific environment due to the limitation of their knowledge and experience, although the experts coming from different fields are viewed as sophisticated scholars. What is more, when the experts make decisions ambiguously, the confusion and imprecision of the outcomes increase with their hesitation.

Therefore, in order to apply this method with complex problem in the uncertain environment, the Fuzzy PROMETHEE was proposed and used in many areas such as material selection (Gul et al. 2018), decision on the information system/information technology outsource (Chen et al. 2011), evaluation of hospital service quality (2016), energy exploitation projects ranking (Goumas and Lygerou 2000), selection of start-up businesses in a public venture capital financing (Chen et al. 2011), and assessment of nuclear medicine imaging devices (Gul et al. 2018). The procedure of Fuzzy PROMETHEE is exhibited as follows (Chen et al. 2011; Gul et al. 2018; Yatsalo et al. 2020):

Step 1: Given the real-world problem, establish the set of feasible alternatives $A = \{a_i, i = 1, 2, \cdots, m\}$ and the set of evaluation criteria $C = \{c_j, j = 1, 2, \cdots, n\}$.
Step 2: Determine linguistic variables and assign relevant fuzzy numbers for these variables, which are used to assess the relative importance of all alternatives over

each criterion. In this paper, triangular fuzzy number that viewed as a special version of trapezoidal fuzzy number, is adopt to describe the decision makers' evaluation on the alternatives. The triangular fuzzy number is mathematically represented as $\widetilde{A} = [l, m, r]$.

Step 3: Form a group of decision makers to assess the preference of all alternatives through pairwise comparisons using linguistic variables to generate the evaluation matrix, as shown in Eq. (9.16).

$$\widetilde{D} = \left[\tilde{a}_{ij}\right]_{m\times n} = \begin{bmatrix} \tilde{a}_{11} & \tilde{a}_{12} & \cdots & \tilde{a}_{1k} \\ \tilde{a}_{21} & \tilde{a}_{22} & \cdots & \tilde{a}_{2k} \\ \cdots & \cdots & \cdots & \cdots \\ \tilde{a}_{m1} & \tilde{a}_{m2} & \cdots & \tilde{a}_{mk} \end{bmatrix} i = 1, 2, \cdots, m; j = 1, 2, \cdots, n \tag{9.16}$$

Step 4: Select a real preference function $f_k(*)$ for each criterion $c_j \subset C$ according to the feature of the relevant criterion.

Step 5: Evaluate the fuzzy outranking relationship between alternative a_i and alternative a_j to generate the fuzzy preference index of alternatives a_i based on the fuzzy preference function values $P_k(\widetilde{a}_i, \widetilde{a}_j)$, as presented by Eq. (9.17).

$$\begin{aligned} \tilde{\pi}(a_i, a_j) &= \sum_{k=1}^{n} P_k(\widetilde{a}_i, \widetilde{a}_j) \\ &= \sum_{k=1}^{n} \widetilde{f}_k(\widetilde{a}_i - \widetilde{a}_j) \quad where \quad a_i, a_j \in A \quad k = 1, 2, \cdots, n \end{aligned} \tag{9.17}$$

Step 6: Calculate the fuzzy leaving flow and entering flow for each alterative by Eqs. (9.18)–(9.19):

$$\text{Fuzzy leaving flow: } \tilde{\phi}^{+}(a_i) = \frac{1}{m-1} \sum_{a_j \neq a_i}^{m} \tilde{\pi}(\widetilde{a}_i, \widetilde{a}_j)\ \widetilde{a}_i, \widetilde{a}_j \in \widetilde{A} \tag{9.18}$$

$$\text{Fuzzy entering flow } \widetilde{a}_i, \widetilde{a}_j \in \widetilde{A}\ \tilde{\phi}^{-}(a_i) = \frac{1}{m-1} \sum_{a_j \neq a_i}^{m} \tilde{\pi}(\widetilde{a}_j, \widetilde{a}_i)\ \widetilde{a}_i, \widetilde{a}_j \in \widetilde{A} \tag{9.19}$$

Step 7: Defuzzify the fuzzy leaving flows and entering flows. In this paper, the defuzzification method of the center of area (COA) (Sugeno 1985) is used to generate the crisp vale of the fuzzy leaving and entering flow for each alternative.

Eventually, the prioritization of the alternatives is yielded based the method of PROMETHEE-I/II.

9.4 Results and Discussions

To mitigate energy supply risks and concurrently alleviate environmental pollution, as well as protect public health, the promotion of renewable energy development is the key to achieve these goals. Therefore, this paper conducted a study on the performance of various kinds of renewable energy technologies using Fuzzy PROMETHEE with consideration of the influencing factors and risks resulting from different aspects. The current renewable energy technologies, involving hydro, wind, solar, biomass, and hydrogen, were assessed in the multi-criteria environment which consists of the aforementioned risks in Sect. 9.2 belonging to four dimensions with twenty-three components.

Five-scale linguistic variables were defined to express the assessment of the potentials for each renewable energy technology with respect to each criterion, and Table 9.3 has presented the predefined linguistic variables and the corresponding fuzzy numbers. The membership functions of these fuzzy numbers have been shown in Fig. 9.1. And then, an expert group was formed to evaluate the relative performance of various renewable energy technologies using the determined linguistic variable to generate a fuzzy evaluation matrix as shown in Table 9.4. In this study, we assumed that the importance of each expert is equivalent to others and each criterion share the same weight so the fuzzy preference function for each renewable energy technology is calculated according to the steps of the Fuzzy PROMETHEE method introduced in Sect. 9.3. The fuzzy preferences of hydropower over others, $P_k(hyproP, *)$, are illustrated in Table 9.5.

After computing the preference functions of all renewable energy resources, the fuzzy preference indices can be generated using the steps mentioned in Sect. 9.3 (as shown in Table 9.6). Then, the corresponding entering and leaving flows can be calculated, and the results are illustrated in Table 9.7. Further, the net flow of each renewable energy technology can be produced as $\phi(hydroP) = 0.0139$, $\phi(wind) = 0.1571$, $\phi(solar) = 0.0738$, $\phi(biomass) = -0.0807$, and $\phi(hydroG) = -0.1464$. Consequently, the prioritization of the five renewable

Table 9.3 Linguistic variables and the corresponding fuzzy numbers

Linguistic variables	Fuzzy numbers
Very Poor (VP)	[0.00, 0.15, 0.30]
Poor (P)	[0.25, 0.35, 0.45]
Fair (F)	[0.35, 0.55, 0.65]
Good (G)	[0.60, 0.75, 0.85]
Best (B)	[0.80, 0.90, 1.00]

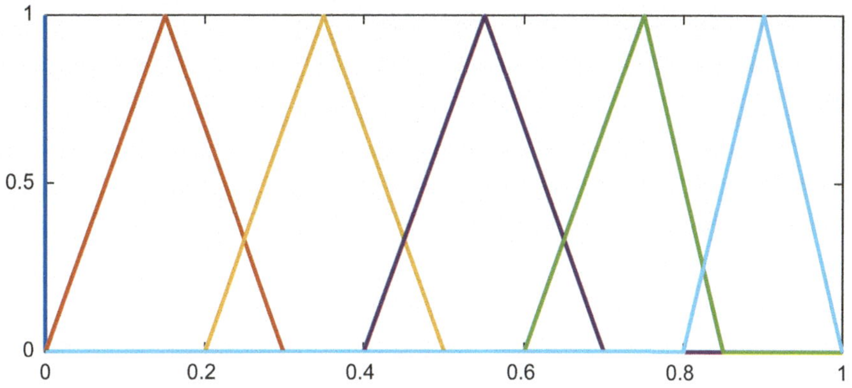

Fig. 9.1 The membership functions used in this study

Table 9.4 Assessment on various renewable energy technologies with respect to each criterion

	hydroP	wind	solar	biomass	hydroG
En1	F	P	F	P	F
En2	G	F	VP	P	VP
En3	B	B	B	F	P
En4	VP	VP	G	VP	F
En5	F	F	P	VP	G
En6	F	VP	F	F	VP
En7	VP	F	F	B	B
Ec1	P	VP	VP	F	G
Ec2	VP	G	P	VP	F
Ec3	G	B	VP	P	F
Ec4	P	F	VP	P	F
Ec5	F	VP	P	B	P
SP1	VP	P	F	VP	F
SP2	G	P	VP	VP	F
SP3	P	B	B	VP	B
SP4	G	F	B	G	F
SP5	G	B	F	F	G
SP6	G	VP	F	G	P
Te1	F	P	P	F	B
Te2	VP	VP	B	F	F
Te3	P	VP	P	P	P
Te4	B	B	VP	G	B
Te5	VP	F	G	VP	B

Table 9.5 The preference function value of hydropower

	$P_k(hyproP, wind)$	$P_k(hyproP, solar)$	$P_k(hyproP, biomass)$	$P_k(hyproP, hydroG)$
En1	(0.2869, 0.6750, 1.1569)	(0.4267, 0.8358, 1.1929)	(0.4698, 0.8664,1.1558)	(0.3010,0.6580, 1.1577)
En2	(0.3189, 0.7539, 1.7387)	(0.3433, 0.6992, 1.3363)	(0.5081, 0.7063, 1.0439)	(0.2948, 0.7296, 1.0748)
En3	(0.1447, 0.7786, 1.1537)	(0.2709, 0.8679, 1.3954)	(0.4750, 0.8191, 1.3016)	(0.3836, 0.8308, 1.4212)
En4	(0.3216, 0.7964, 1.7123)	(0.5476, 0.6959, 1.2140)	(0.5471, 0.7876, 1.3872)	(0.4189, 0.6729, 1.3096)
En5	(0.2227, 0.6084, 1.1429)	(0.4670, 0.8435, 1.3681)	(0.5384, 0.7452, 1.2060)	(0.3189, 0.8806, 1.3157)
En6	(0.1343, 0.7670, 1.8270)	(0.3115, 1.1748, 1.5054)	(0.4376, 0.7501, 1.0363)	(0.4061, 0.6923, 1.4600)
En7	(0.1981, 0.7573, 1.2491)	(0.2395, 0.6755, 1.4528)	(0.4795, 0.8184, 1.1259)	(0.3851, 0.9486, 1.5369)
Ec1	(0.1926, 0.5639, 1.0961)	(0.2424, 0.7282, 1.2736)	(0.3546, 0.7264, 1.0684)	(0.2513, 0.7022, 1.2387)
Ec2	(0.3371, 0.5467, 1.1505)	(0.3007, 0.6928, 1.4674)	(0.6452, 0.8676, 1.1333)	(0.3704, 0.8909, 1.4600)
Ec3	(0.3397, 0.6958, 1.5146)	(0.3768, 0.5805, 1.4131)	(0.4002, 0.8319, 1.2144)	(0.3274, 0.6717, 1.2855)
Ec4	(0.1555, 0.8770, 1.3722)	(0.2700, 0.7950, 1.4408)	(0.3819, 0.8949, 1.2588)	(0.4332, 0.7092, 1.5878)
Ec5	(0.3417, 0.6337, 1.2508)	(0.2296, 0.7819, 1.3085)	(0.4617, 0.8960, 1.2233)	(0.2502, 0.6346, 1.5814)
SP1	(0.3370, 0.7299, 1.7288)	(0.3592, 1.1599, 1.3650)	(0.4094, 0.8788, 1.4364)	(0.3425, 0.8190, 1.1721)
SP2	(0.1709, 0.5879, 1.4839)	(0.4871, 0.8571, 1.4062)	(0.4969, 0.7094, 1.2542)	(0.3349, 0.8597, 1.4256)
SP3	(0.2818, 0.7951, 1.4484)	(0.4001, 0.6834, 1.1899)	(0.4518, 0.7193, 1.4712)	(0.3422, 0.8077, 1.4034)
SP4	(0.1249, 0.6002, 1.8150)	(0.4039, 1.0624, 1.4404)	(0.6355, 0.7895, 1.3152)	(0.4040, 0.7618, 1.3739)
SP5	(0.1485, 0.6987, 1.1852)	(0.3346, 1.1091, 1.5935)	(0.6261, 0.8486, 1.4784)	(0.3145, 0.8449, 1.6077)
SP6	(0.3224, 0.7746, 1.6554)	(0.4681, 0.7949, 1.3856)	(0.3658, 0.8832, 1.1128)	(0.4069, 0.8461, 1.1780)
Te1	(0.2789, 0.8499, 1.6519)	(0.3177, 0.5362, 1.2295)	(0.5714, 0.8141, 1.3348)	(0.3443, 0.8580, 1.2158)
Te2	(0.3378, 0.8768, 1.2795)	(0.3557, 0.8605, 1.2804)	(0.4307, 0.9296, 1.1374)	(0.2572, 0.8416, 1.0746)

(continued)

Table 9.5 (continued)

	$P_k(hyproP, wind)$	$P_k(hyproP, solar)$	$P_k(hyproP, biomass)$	$P_k(hyproP, hydroG)$
Te3	(0.2309, 0.7149, 1.4665)	(0.3406, 0.8929, 1.3937)	(0.4769, 0.9011, 1.3326)	(0.2852, 0.9592, 1.6573)
Te4	(0.1260, 0.5545, 1.6757)	(0.3842, 0.8836, 1.3115)	(0.5144, 0.9712, 1.3445)	(0.3944, 0.6794, 1.4483)
Te5	(0.2990, 0.5586, 1.2538)	(0.2672, 0.8120, 1.4174)	(0.6328, 0.8488, 1.0247)	(0.3447, 0.8695, 1.3304)

Table 9.6 The fuzzy preference indices for all renewable energy technologies

$\tilde{\pi}(hydroP, wind)$	(0.2375, 0.7041, 1.4352)	$\tilde{\pi}(biomass, hydroP)$	(0.3207, 0.6839, 1.3819)
$\tilde{\pi}(hydroP, solar)$	(0.3541, 0.8271, 1.3644)	$\tilde{\pi}(biomass, wind)$	(0.3068, 0.7792, 1.2980)
$\tilde{\pi}(hydroP, biomass)$	(0.4918, 0.8262, 1.2347)	$\tilde{\pi}(biomass, solar)$	(0.3931, 0.7106, 1.1713)
$\tilde{\pi}(hydroP, hydroG)$	(0.3440, 0.7899, 1.3616)	$\tilde{\pi}(biomass, hydroG)$	(0.5666, 0.8738, 1.2583)
$\tilde{\pi}(wind, hydroP)$	(0.4014, 0.7942, 1.6106)	$\tilde{\pi}(hydroG, hydroP)$	(0.3071, 0.5663, 1.3646)
$\tilde{\pi}(wind, solar)$	(0.3683, 0.7830, 1.3287)	$\tilde{\pi}(hydroG, wind)$	(0.2955, 0.5631, 1.3150
$\tilde{\pi}(wind, biomass)$	(0.4807, 0.9394, 1.5207)	$\tilde{\pi}(hydroG, biomass)$	(0.3079, 0.7666, 1.1502)
$\tilde{\pi}(wind, hydroG)$	(0.4204, 0.9293, 1.6720)	$\tilde{\pi}(hydroG, solar)$	(0.3096, 0.8662, 1.2606)
$\tilde{\pi}(solar, hydroP)$	(0.2937, 0.6726, 1.4070)		
$\tilde{\pi}(solar, wind)$	(0.3320, 0.7570, 1.3401)		
$\tilde{\pi}(solar, biomass)$	(0.4549, 0.9534, 1.5865)		
$\tilde{\pi}(solar, hydroG)$	(0.3218, 0.8593, 1.4326)		

Table 9.7 The leaving and entering flows for all renewable energy technologies

	$\tilde{\phi}^+(*)$	$\tilde{\phi}^+(*)$	$\tilde{\phi}^-(*)$	$\tilde{\phi}^-(*)$
hydroP	0.3569 0.7868 1.3490	0.8309	0.3307 0.6793 1.4410	0.8170
wind	0.4177 0.8615 1.5330	0.9374	0.2930 0.7009 1.3471	0.7803
solar	0.3506 0.8106 1.4416	0.8676	0.3559 0.7718 1.2537	0.7938
biomass	0.3968 0.7619 1.2774	0.8120	0.4338 0.8714 1.3730	0.8928
hydroG	0.3050 0.6906 1.2726	0.7561	0.4132 0.8631 1.4311	0.9025

energy options is wind power, solar energy, hydropower, biomass, and hydrogen energy according to the PROMETHEE-II method. Figure 9.2 shows the ranking relation between these alternative pairs.

The performance of various renewable energy technologies has been assessed by using Fuzzy PROMETHEE with the comprehensive consideration of a series of environmental, economic, social-political, and technological risks and factors in this paper. The results demonstrate that wind power has achieved a higher performance, which imply its greater potential in China than other renewable energy options.

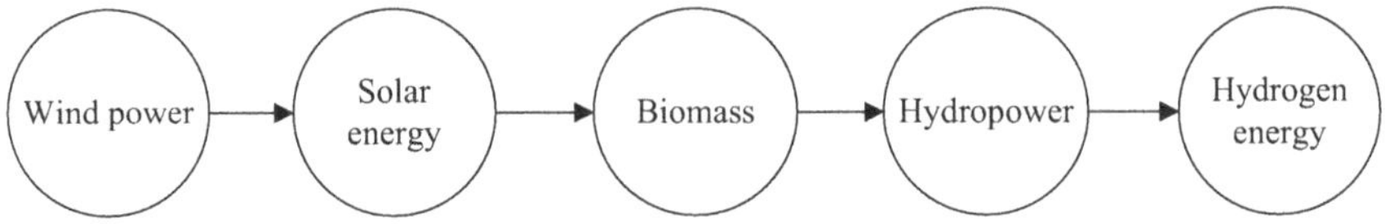

Fig. 9.2 Priority of different renewable energy technologies

Compared with other technologies, wind power plants are often located in the wilderness far away from residential areas, so it has very little impact on the daily life of residents. Therefore, the residents have a relatively higher degree of acceptance to it, despite its impact on landscape. Moreover, wind power technology is relatively mature, and there are broad offshore coasts for developing offshore wind power in China.

The inferior technology is solar, as the results indicate. Presently, this technology has been more developed and economically feasible than a decade ago. However, its social acceptance seems to be a major risk for its development. The residents are convinced that the promotion of solar photovoltaic projects may have adverse risks on their living environment and believe the operation of the devices also more or less affects their health. For instance, the installation and operation of photovoltaic projects may affect local biodiversity and generate light pollution to residents. In addition, the lack of adequate information on photovoltaic devices also leads to cognitive bias as well as the overstatement of the negative effects in the production of solar energy, especially in the rural area.

As to biomass, although there are a lot of material sources for biomass production, such as agricultural fruits and straws, wastes food and cooking oil, animal wastes, and municipal solid wastes, most of them have a very low energy density, and a convenient and efficient supply chain network is required, which make it very expensive. Agricultural fruits may be an ideal material source for bioenergy production, while the growth of biomass material needs to occupy precious agricultural land and water resources, which results in competition between energy production and food supply due to the complex and difficult nexus among energy, food, and water resource.

As to hydropower, it has been viewed as the conventional renewable energy source in China, while the relative potential is no higher than the aforementioned sources. A major reason is that the construction of hydropower plants is extensively dependent on the natural conditions. Besides, hydropower has already accounted for a large portion of the total production of renewable energy in China, and the central government has decided to stop the development of hydropower in some regions for protecting biodiversity. So it is necessary to develop other renewable energy sources for the purpose of enhancing the diversity of renewable energy supply and strengthening energy security.

Eventually, regarding the novel renewable energy of hydrogen, its production and utilization are faced up with various difficult issues, such as the high cost for hydrogen production and storage, safety issues in hydrogen transportation and utilization, the establishment of hydrogen refueling stations, and the corresponding mobile device in innovation. Accordingly, the relative importance of hydrogen is inferior to other forms of renewable energy.

It is worth noting that the results are just produced based on the consideration of existing technical conditions and market environment. In fact, public awareness and attitude toward renewable energy technologies can also be potential risks and threats for renewable energy development.

9.5 Conclusions

The urgent need for mitigating energy supply risks and transiting to sustainable development incentivize the increasing interests in the researches related to renewable energy. This paper reviewed the existing literature on the risks influencing renewable energy development and constructed a shared framework for further analyzing them. What is more, in order to draw an alternative proper developing path for renewable energy in China, we assessed the potential of the current renewable energy technologies under the built framework by forming an expert group and using a multicriteria decision analysis method, fuzzy PROMETHEE. It is concluded that wind power is superior to other renewables, and solar energy and biomass outperform hydropower and hydrogen energy based on the consideration of environmental, economic, social-political, and technological factors.

In order to promote the utilization of solar photovoltaic and biomass, the focus academic research is suggested to shift to the social components gradually from the technological risks. Especially, to facilitate renewable energy development, it is very important for the government to win public support to renewable energy projects. In order to do that, the government should timely disclosure the official information about the novel technologies and actively publicize the construction of renewable energy projects at the early stage. As for hydropower as a traditional form of renewable energy around the world, it is heavily dependent on the geographical conditions, even poses threats to local biodiversity and increases risks of geologic hazards. Thus, the hydropower development should be conducted judiciously and cautiously, and the expansion speed of traditional plants should be controlled to a proper range.

Hydrogen energy, especially green hydrogen, has been taken as a promising energy source for sustainable energy system; more investment in the research and development is to do the aim of ensuring the safety and feasibility of its production, transportation, and utilization. Mitigating energy security risks not only depends on the promotion of renewable energy but heavily relies on a diversity of renewable energy sources and various innovative energy technologies. It is also of great significance to create an informative and supportive social environment to facilitate renewable energy development.

References

Aaen SB, Kerndrup S, Lyhne I (2016) Beyond public acceptance of energy infrastructure: How citizens make sense and form reactions by enacting networks of entities in infrastructure development. Energy Policy 96:576–586

Ackermann T, Söder L (2002) An overview of wind energy-status 2002. Renew Sustain Energy Rev 6(1):67–127

Afful-Dadzie E, Oplatková ZK, Nabareseh S (2015) Selecting Start-Up Businesses in a Public Venture Capital Financing using Fuzzy PROMETHEE. Procedia Computer Science 60:63–72

Agterbosch S, Glasbergen P, Vermeulen WJV (2007) Social barriers in wind power implementation in The Netherlands: Perceptions of wind power entrepreneurs and local civil servants of institutional and social conditions in realizing wind power projects. Renew Sustain Energy Rev 11(6):1025–1055

Aitken M (2010) Why we still don't understand the social aspects of wind power: A critique of key assumptions within the literature. Energy Policy 38(4):1834–1841

Asadullah M (2014) Barriers of commercial power generation using biomass gasification gas: A review. Renew Sustain Energy Rev 29:201–215

Bell D, Gray T, Haggett C (2005) The 'Social Gap' in Wind Farm Siting Decisions: Explanations and Policy Responses. Environmental Politics 14(4):460–477

Bertsch V et al (2016) Public acceptance and preferences related to renewable energy and grid expansion policy: Empirical insights for Germany. Energy 114:465–477

Bob D (2015) BP statistical review of world energy. London, UK

Brans JP, Vincle P (1985) A preference ranking organization method. Manage Sci 31:647–656

Brans JP, Vincke P, Marescha B (1986) How to select and how to rank projects: the PROMETHEE method. Eur J Oper Res 24:228–238

Brennan N, Van Rensburg TM (2016) Wind farm externalities and public preferences for community consultation in Ireland: A discrete choice experiments approach. Energy Policy 94:355–365

Breukers S, Wolsink M (2007) Wind power implementation in changing institutional landscapes: An international comparison. Energy Policy 35(5):2737–2750

Bright J et al (2008) Map of bird sensitivities to wind farms in Scotland: A tool to aid planning and conservation. Biol Cons 141(9):2342–2356

Campbell R et al (2018) Technoeconomic and Policy Drivers of Project Performance for Bioenergy Alternatives Using Biomass from Beetle-Killed Trees. Energies 11(2):293

Capodaglio AG, Callegari A, Lopez MV (2016) European framework for the diffusion of biogas uses: Emerging technologies, acceptance, incentive strategies, and institutional-regulatory support. Sustainability 8(4):1–18

Chen Y-H, Wang T-C, Wu C-Y (2011) Strategic decisions using the fuzzy PROMETHEE for IS outsourcing. Expert Syst Appl 38(10):13216–13222

Chiras D, Sagrillo M, Woofenden I (2009) Power from the wind: achieving energy independence, 1st edn. New Society Publishers, Canada

Cicia G et al (2012) Fossil energy versus nuclear, wind, solar and agricultural biomass: Insights from an Italian national survey. Energy Policy 42:59–66

Çoban V, Onar SÇ (2017) Modeling renewable energy usage with hesitant Fuzzy cognitive map. Complex & Intelligent Systems 3(3):155–166

Corwin S, Johnson TL (2019) The role of local governments in the development of China's solar photovoltaic industry. Energy Policy 130:283–293

de Sena LA, Ferreira P, Braga AC (2016) Social acceptance of wind and solar power in the Brazilian electricity system. Environ Dev Sustain 18(5):1457–1476

Demski C, Poortinga W, Pidgeon N (2014) Exploring public perceptions of energy security risks in the UK. Energy Policy 66:369–378

Devine-Wright P (2007) Reconsidering public attitudes and public acceptance of renewable energy technologies: a critical review. Beyond Nimbyism: a multidisciplinary investigation of public engagement with renewable energy technologies: 15

Devine-Wright P (2005) Beyond NIMBYism: towards an integrated framework for understanding public perceptions of wind energy. Wind Energy: an International Journal for Progress and Applications in Wind Power Conversion Technology 8(2):125–139
Devine-Wright P (2013) Explaining "NIMBY" objections to a power line: The role of personal, place attachment and project-related factors. Environ Behav 45(6):761–781
Echegaray F (2014) Understanding stakeholders' views and support for solar energy in Brazil. J Clean Prod 63:125–133
Eleftheriadis IM, Anagnostopoulou EG (2015) Identifying barriers in the diffusion of renewable energy sources. Energy Policy 80:153–164
Ertör-Akyazı P et al (2012) Citizens' preferences on nuclear and renewable energy sources: Evidence from Turkey. Energy Policy 47:309–320
Feng Y et al (2015) Overview of wind power generation in China: Status and development. Renew Sustain Energy Rev 50:847–858
Firestone J et al (2012) Public Acceptance of Offshore Wind Power across Regions and through Time. JOURNAL OF ENVIRONMENTAL PLANNING AND MANAGEMENT 55(10):1369–1386
Frantál B (2015) Have Local Government and Public Expectations of Wind Energy Project Benefits Been Met? Implications for Repowering Schemes. J Environ Planning Policy Manage 17(2):217–236
Ghimire LP, Kim Y (2018) An analysis on barriers to renewable energy development in the context of Nepal using AHP. Renewable Energy 129:446–456
Goumas M, Lygerou V (2000) An extension of the PROMETHEE method for decision making in fuzzy environment: Ranking of alternative energy exploitation projects. Eur J Oper Res 123(3):606–613
Greenberg M (2009) Energy sources, public policy, and public preferences: Analysis of US national and site-specific data. Energy Policy 37(8):3242–3249
Gul M et al (2018) A fuzzy logic based PROMETHEE method for material selection problems. Beni-Suef University Journal of Basic and Applied Sciences 7(1):68–79
Gul M et al (2018) A fuzzy logic based PROMETHEE method for material selection problems. Beni-Suef University Journal of Basic and Applied Sci 7(1): p. 68–79
He Z et al (2016) Overview of the development of the Chinese Jiangsu coastal wind-power industry cluster. Renew Sustain Energy Rev 57:59–71
Hernandez RR et al (2014) Environmental impacts of utility-scale solar energy. Renew Sustain Energy Rev 29:766–779
Hobman EV, Ashworth P (2013) Public support for energy sources and related technologies: The impact of simple information provision. Energy Policy 63:862–869
Hu Y et al (2016) Current status, challenges, and perspectives of Sichuan's renewable energy development in Southwest China. Renew Sustain Energy Rev 57:1373–1385
Huang S-C, Lo S-L, Lin Y-C (2013) Application of a fuzzy cognitive map based on a structural equation model for the identification of limitations to the development of wind power. Energy Policy 63:851–861
Jacobsson S, Johnson A (2000) The diffusion of renewable energy technology: an analytical framework and key issues for research. Energy Policy 28(9):625–640
Jagadeesh A (2000) Wind energy development in Tamil Nadu and Andhra Pradesh, India Institutional dynamics and barriers — A case study. Energy Policy 28(3):157–168
Jobert A, Laborgne P, Mimler S (2007) Local acceptance of wind energy: Factors of success identified in French and German case studies. Energy Policy 35(5):2751–2760
Kermagoret C et al (2016) Stakeholder Perceptions of Offshore Wind Power: A Fuzzy Cognitive Mapping Approach. Soc Nat Resour 29(8):916–931
Komarek TM, Lupi F, Kaplowitz MD (2011) Valuing energy policy attributes for environmental management: Choice experiment evidence from a research institution. Energy Policy 39(9):5105–5115

Ladenburg J, Möller B (2011) Attitude and acceptance of offshore wind farms—The influence of travel time and wind farm attributes. Renew Sustain Energy Rev 15(9):4223–4235
Langer K et al (2016) A qualitative analysis to understand the acceptance of wind energy in Bavaria. Renew Sustain Energy Rev 64:248–259
Li et al (2014) Latest development status of offshore wind power in China—The perspective of developers. Journal of Renewable Sustainable Energy & Fuels 6(5): p. 053138
Liu MFM, Pistorius T (2012) Coping with the energy crisis: Impact assessment and potentials of non-traditional renewable energy in rural Kyrgyzstan. Energy Policy 44:130–139
Luthra S et al (2015) Barriers to renewable/sustainable energy technologies adoption: Indian perspective. Renew Sustain Energy Rev 41:762–776
Mallett A (2007) Social acceptance of renewable energy innovations: The role of technology cooperation in urban Mexico. Energy Policy 35(5):2790–2798
Mann J, Teilmann J (2013) Environmental impact of wind energy. Environmental Research Lett 8(3): p. 035001.
Manwell JF, McGowan JG, Rogers AL (2010) Wind energy explained: theory, design and application. Wiley
Mercer N, Sabau G, Klinke A (2017) "Wind energy is not an issue for government": Barriers to wind energy development in Newfoundland and Labrador, Canada. Energy Policy 108:673–683
Mittal S, Ahlgren EO, Shukla PR (2018) Barriers to biogas dissemination in India: A review. Energy Policy 112:361–370
Nadaï A, Labussière O (2009) Wind power planning in France (Aveyron), from state regulation to local planning. Land Use Policy 26(3):744–754
Ozonoh M et al (2018) Techno-economic analysis of electricity and heat production by co-gasification of coal, biomass and waste tyre in South Africa. J Clean Prod 201:192–206
Ozsahin DU et al (2017) Evaluating nuclear medicine imaging devices using fuzzy PROMETHEE method. Procedia Computer Science 120:699–705
Painuly JP (2001) Barriers to renewable energy penetration; a framework for analysis. Renewable Energy 24(1):73–89
Rahma MM et al (2017) Environmental Impact Assessment of Different Renewable Energy Res 29–71
Reddy S, Painuly JP (2004) Diffusion of renewable energy technologies—barriers and stakeholders' perspectives. Renewable Energy 29(9):1431–1447
Ren J, Sovacool BK (2015) Prioritizing low-carbon energy sources to enhance China's energy security. Energy Convers Manage 92:129–136
Ribeiro F, Ferreira P, Araújo M (2011) The inclusion of social aspects in power planning. Renew Sustain Energy Rev 15(9):4361–4369
Ribeiro F et al (2014) Public opinion on renewable energy technologies in Portugal. Energy 69:39–50
Richards G, Noble B, Belcher K (2012) Barriers to renewable energy development: A case study of large-scale wind energy in Saskatchewan, Canada. Energy Policy 42:691–698
Roques F, Hiroux C, Saguan M (2010) Optimal wind power deployment in Europe—A portfolio approach. Energy Policy 38(7):3245–3256
Saidur R et al (2011) Environmental impact of wind energy. Renew Sustain Energy Rev 15(5):2423–2430
Sennarogl B, Tuzkaya MMG (2016) An interval valued intuitionistic fuzzy PROMETHEE approach for hospital service quality evaluation. in Uncertainty Modelling in Knowledge Engineering and Decision Making: Proceedings of the 12th International FLINS Conference. 2016. World Scientific
Shaaban M, Petinrin JO (2014) Renewable energy potentials in Nigeria: Meeting rural energy needs. Renew Sustain Energy Rev 29:72–84
Soland M, Steimer N, Walter G (2013) Local acceptance of existing biogas plants in Switzerland. Energy Policy 61:802–810
Sovacool BK (2009) Rejecting renewables: The socio-technical impediments to renewable electricity in the United States. Energy Policy 37(11):4500–4513

Sovacool BK, Blyth PL (2015) Energy and environmental attitudes in the green state of Denmark: Implications for energy democracy, low carbon transitions, and energy literacy. Environ Sci Policy 54:304–315
Sovacool BK, Blyth PL (2015) Energy and environmental attitudes in the green state of Denmark: Implications for energy democracy, low carbon transitions, and energy literacy. Environ Sci Policy 54:304–315
Sugeno M (1985) An introductory survey of fuzzy control. Inf Sci 36(1–2):59–83

Upham P et al (2009) Public attitudes to environmental change: a selective review of theory and practice. A research synthesis for the Living with Environmental Change Programme. In: Living wiht environmental change LWEC. Research Councils UK: UK

van Rensburg TM, Kelley H, Jeserich N (2015) What influences the probability of wind farm planning approval: Evidence from Ireland. Ecol Econ 111:12–22

Warren CR et al (2005) 'Green On Green': Public perceptions of wind power in Scotland and Ireland. J Environ Planning Manage 48(6):853–875

Wizelius T (2015) Developing wind power projects: theory and practice. Routledge

Wolsink M (2000) Wind power and the NIMBY-myth: institutional capacity and the limited significance of public support. Renewable Energy 21(1):49–64

Wolsink M (2007) Wind power implementation: The nature of public attitudes: Equity and fairness instead of 'backyard motives.' Renew Sustain Energy Rev 11(6):1188–1207

Wolsink M (2010) Contested environmental policy infrastructure: Socio-political acceptance of renewable energy, water, and waste facilities. Environ Impact Assess Rev 30(5):302–311

Wüstenhagen R, Wolsink M, Bürer MJ (2007) Social acceptance of renewable energy innovation: An introduction to the concept. Energy Policy 35(5):2683–2691

Kunze C, Busch H (2011) The social complexity of renewable energy production in the countryside. Electronic Green J 1(31)

Yatsalo B et al (2020) Fuzzy extensions of PROMETHEE: Models of different complexity with different ranking methods and their comparison. Fuzzy Sets and Systems, 2020.

Yuan X, Zuo J, Ma C (2011) Social acceptance of solar energy technologies in China—End users' perspective. Energy Policy 39(3):1031–1036

Zadeh LA (1965) Fuzzy sets. Inf Control 8(3):338–353

Zeng M et al (2015) Orderly grid connection of renewable energy generation in China: Management mode, existing problems and solutions. Renew Sustain Energy Rev 41:14–28

Zhang X, Shen L, Chan SY (2012) The diffusion of solar energy use in HK: What are the barriers? Energy Policy 41:241–249

Zhang Z et al (2013) Barriers to commercialization development of crop straw gasification technology in China and promoting policy design. Energy Sources Part B 8(3):279–289

Zhang Y et al (2016) Wind energy rejection in China: Current status, reasons and perspectives. Renew Sustain Energy Rev 66:322–344

Zhang L et al (2017) Measuring energy security performance within China: Toward an inter-provincial prospective. Energy 125:825–836

Chapter 10
Sustainability Risks of Resource-Exhausted Cities in China: A Principal Component Analysis

Huijuan Xiao, Long Zhang, and Jingzheng Ren

Abstract Resource-exhausted cities indicate the cities whose natural resources are exhausting, and whose accumulated exploitation reserves have reached more than 70% of the recoverable reserves. These types of cities could encounter many sustainability risks caused by the resource curse, such as slowing economic development, difficulty in economic transformation, sluggish in fostering new growth points, rising unemployment, insufficient innovation capacity, and deterioration of environmental and ecological systems. However, the status of sustainability of resource-exhausted cities is still unclear. To fill this research gap, the study constructs an evaluation framework covering economic, social, and environmental dimensions for the sustainability of resource-exhausted cities. The principal component analysis is used to evaluate the sustainability of resource-exhausted cities in China from 2005 to 2016. Results show that (1) the sustainability of resource-exhausted cities sees an increasing trend. However, the average sustainability of these cities is 64.297 in 2016 and there is still much room for improvement; (2) Shuangyashan, Fushun, Fuxin, Shizuishan, and Wuhai see the most significant risk in sustainability, at a mere 52.242, 52.447, 47.371, 34.062, and 4.113 in 2016, respectively; and (3) Panjin, Shuangyashan, and Yichun-HLJ are the only three cities whose sustainability sees a decreasing trend from 2005 to 2016. This indicates that Fuxin, Liaoyuan, and Shuangyashan face relatively significant sustainability risks and prompt actions should be taken to reverse this deterioration. The results obtained in this study can be a reference to take stock of where resource-exhausted cities stand in terms of sustainability, identify the potential risks, and further promote sustainable development.

Keywords Risk · Resource-exhausted cities · Sustainability · Principal component analysis · Resource curse

---

H. Xiao · L. Zhang · J. Ren (✉)
Department of Industrial and Systems Engineering, The Hong Kong Polytechnic University, Hong Kong, China
e-mail: jingzheng.jz.ren@polyu.edu.hk

J. Ren (ed.), *Multi-Criteria Decision Analysis for Risk Assessment and Management*, Industrial Ecology and Environmental Management 1,
https://doi.org/10.1007/978-3-030-78152-1_10

10.1 Introduction

Sustainable development is the development that meets the needs of the present generation without compromising the ability of future generations to meet their own needs, which is defined by the Brundtland commission in 1987 (Brundtland et al. 1987). Here, we define sustainability risks as the factors that can pose risks to achieve a better and more sustainable future of a given system (e.g., an economy, a corporation, and a community). These sustainability risks can be related to economy, environment, and society. Resource-exhausted cities indicate the cities whose natural resources are exhausting and are in a later, late, or even end stages after continuous exploitation, and whose accumulated exploitation reserves have reached more than 70% of the recoverable reserves. China is not the only country that has resource-exhausted cities, in fact, these types of cities exist in many countries around the world. According to the National Development and Reform Commission of China, 69 resource-exhausted cities have been identified in this country (The Chinese Government 2011). These cities may encounter many sustainability risks caused by the resource curse, which have attracted great attention from many scholars and policy-makers (The Chinese Government 2008; Zhang et al. 2018).

The term 'curse of natural resource' is proposed by Auty et al. (1998) denoting the paradoxical situation that the development of resource-rich regions falls behind those with relatively poor natural resources. Some studies have found that the resource curse happens in China, especially in the central and western regions (Shao and Qi 2009; Zhang and Brouwer 2020). Shao and Qi (2009) suggested that there was a negative relationship between economic development and energy exploitation in Western China since the 1990s. The sustainability risks of resource-exhausted cities can be closely related to the negative impacts exerted by the resource curse. The transmission effects of the resource curse can be categorized into three main mechanisms (Shao et al. 2020; Szalai 2018), that is Dutch disease, crowding-out effect, and institutional weakening effect.

With China's economic restructuring and fundamental change of supply–demand in the resources market, resource-exhausted cities of China are faced with a series of dilemmas regarding the economy, society, and environment (Dong et al. 2007; He et al. 2017; Li et al. 2020). The dilemmas include slowing economic development, difficulty in economic transformation, sluggish in fostering new growth points, rising unemployment, insufficient innovation capacity, institutional issues, and deterioration of environmental and ecological systems. However, the sustainability risks of resource-exhausted cities are still unclear, even though numerous studies have examined these risks for Chinese resource-based cities (Lu et al. 2016; Qin et al. 2019).

To fill this research gap, the study constructs a framework covering economic, social, and environmental dimensions for the sustainability evaluation of resource-exhausted cities in China. The sustainability risks of resource-exhausted cities are multidimensional and complicated. The principal component analysis (PCA) is capable of summarizing the information of large dimensions to smaller dimensions while retaining the data information to a maximum degree (Liou et al. 2004; Vega et al.

1998). As such, the study uses the PCA method to evaluate the sustainability risks of resource-exhausted cities in China from 2005 to 2016. The results can be a reference to take stock of where resource-exhausted cities stand in terms of sustainability, identify the potential risks, and further promote sustainable development.

10.2 Methodology and Data

10.2.1 *The Principal Component Analysis*

The PCA conducts dimension reduction by discarding the highly correlated data information and generates irrelevant components. Thus, PCA can be regarded as one of the most widely adopted statistical tools to reduce dimensions and improve working efficiency for the dataset with many indicators. Considering these advantages of PCA, it has been widely applied in many study areas for performance evaluation and ranking (Omrani et al. 2019; Zhu 1998). In this study, PCA is used to extract information from the dataset with 14 indicators and reduce the dimensions to 7. The procedure of PCA used to evaluate the sustainability risk is provided and some steps are based on Zhu (1998).

Step 1 (Constructing the origin matrix $X = \left[x_{ij}\right]_{n\times p}$). Suppose the dataset has n decision-making unit (DMU) and p indicators. The origin matrix can be constructed as follows.

$$X = \begin{bmatrix} x_{11} & x_{12} & \cdots & x_{1p} \\ x_{21} & x_{22} & \cdots & x_{2p} \\ \vdots & \vdots & \ddots & \vdots \\ x_{n1} & x_{n2} & \cdots & x_{np} \end{bmatrix} \tag{10.1}$$

where x_{ij} is the value of indicator j under DMU i.

Step 2 (Transform all indicators to positive dimension indicator and obtain $Y = \left[y_{ij}\right]_{n\times p}$). The indicators in negative dimension mean the increase of the value of these indicators could deteriorate the performance of DMUs. For consistency, we need to transform the indicators in negative dimension to positive dimension, as follows.

$$y_{ij} = \begin{cases} x_{ij}, & \text{for indicators in positive dimension} \\ -x_{ij}, & \text{for indicators in negative dimension} \end{cases} \tag{10.2}$$

Step 3 (Standardize the matrix Y and obtain standardized matrix $Z = \left[z_{ij}\right]_{n\times p}$). Without loss of generality, all the variables should be standardized to ensure each of them has sample mean 0 and variance of 1. The standardization formula is as follows.

$$z_{ij} = \frac{y_{ij} - \overline{y}_j}{s_j} \quad i = 1, 2, \ldots, n \quad and \quad j = 1, 2, \ldots, p \tag{10.3}$$

$$Z = \left[z_{ij}\right]_{n\times p} = \begin{bmatrix} z_{11} & z_{12} & \cdots & z_{1p} \\ z_{21} & z_{22} & \cdots & z_{2p} \\ \vdots & \vdots & \ddots & \vdots \\ z_{n1} & z_{n2} & \cdots & z_{np} \end{bmatrix} \tag{10.4}$$

$\overline{y}$ and s_j indicate the average value $\left(\frac{\sum_{i=1}^{n} y_{ij}}{n}\right)$ and standard deviation $\left(\sqrt{\frac{\sum_{i=1}^{n}(y_{ij}-\overline{y}_j)^2}{n-1}}\right)$ of indicator j.

Step 4 (Construct the covariance matrix $R = \left[r_{ij}\right]_{p\times p}$). The covariance matrix among the p indicators can be expressed as follows:

$$R = \left[r_{ij}\right]_{p\times p} = \frac{Z^T Z}{n-1} = \begin{bmatrix} \frac{1}{n-1}\sum_{i=1}^{p}(z_{i1})^2 & \frac{1}{n-1}\sum_{i=1}^{p} z_{i1}z_{i2} & \cdots & \frac{1}{n-1}\sum_{i=1}^{p} z_{i1}z_{ip} \\ \frac{1}{n-1}\sum_{i=1}^{p} z_{i2}z_{i1} & \frac{1}{n-1}\sum_{i=1}^{p}(z_{i2})^2 & \cdots & \frac{1}{n-1}\sum_{i=1}^{p} z_{i2}z_{ip} \\ \vdots & \vdots & \ddots & \vdots \\ \frac{1}{n-1}\sum_{i=1}^{p} z_{ip}z_{i1} & \frac{1}{n-1}\sum_{i=1}^{p} z_{ip}z_{i2} & \cdots & \frac{1}{n-1}\sum_{i=1}^{p}(z_{ip})^2 \end{bmatrix} \tag{10.5}$$

Matrix Z^T is the transpose of the matrix Z. The diagonal elements of the matrix Z indicate the variance of a specific indicator, while other elements mean the covariance between two indicators. If the covariance between indicators is high, this dataset is more suitable for conducting PCA. To be noticed, the covariance matrix Z is equal to its correlation matrix using the Pearson correlation method since the matrix Z has been standardized in step 3. To be specific, the formula of correlation between indicator a and indicator b is $\rho_{a,b} = \frac{COV(a,b)}{\sigma_a\sigma_b}$ where $\sigma_a = 1$ and $\sigma_b = 1$. Therefore, $\rho_{a,b} = \frac{COV(a,b)}{\sigma_a\sigma_b} = COV(a,b)$.

Step 5 (Calculate the eigenvalue (λ_k) and eigenvector ($\vec{b_k}$)). k indicates the newly constructed components. The number of the components is the same as the number of indicator j. The eigenvalue of the component k is denoted as λ_k, while the eigenvector is denoted as b_k. According to the properties of the matrix, the number of eigenvalues of a matrix is the same as its order. The eigenvalues can be the same (the multiple roots). As such, we can obtain p eigenvalues and $\lambda_1 \geq \lambda_1 \geq \cdots \geq \lambda_p \geq 0$, the calculation process of the eigenvalue is as follows.

$$\left|R - \lambda I_p\right| = 0 \tag{10.6}$$

where I_p is the identity matrix of size p. The eigenvector ($\vec{b}_k$) corresponding to the kth eigenvalue (λ_k) can be obtained as follows.

$$R\vec{b}_k = \lambda_k \vec{b}_k,\ \ \vec{b}_k = \begin{bmatrix} b_{1k} \\ b_{2k} \\ \vdots \\ b_{pk} \end{bmatrix},\qquad k = 1, 2, \ldots, p \tag{10.7}$$

Since the covariance matrix R is symmetric, these p eigenvectors are perpendicular and not correlated with each other. Then we can obtain a matrix $B = \left[b_{jk}\right]_{p\times p}$ containing p eigenvectors as follows.

$$B = \left[b_{jk}\right]_{p\times p} = \left[\vec{b}_1, \vec{b}_2, \cdots, \vec{b}_p\right] = \begin{bmatrix} b_{11} & b_{12} & \cdots & b_{1p} \\ b_{21} & b_{22} & \cdots & b_{2p} \\ \vdots & \vdots & \ddots & \vdots \\ b_{p1} & b_{p2} & \cdots & b_{pp} \end{bmatrix},\quad j = 1, 2, \ldots, p,\quad k = 1, 2, \ldots, p \tag{10.8}$$

Step 6 (determine the number of principal components to be considered). The number of the components of the dataset is equal to the number of the eigenvalues/the eigenvectors/the indicators (p). The eigenvalues can be the same (the multiple roots), and the eigenvectors obtained based on the multiple roots are also irrelevant since the covariance matrix R is a real symmetric matrix. We choose the top m components as the principal components based on the descending order of the eigenvalue that can cumulatively express more than 80% of data variance. In other words, more than 80% of the information of the dataset can be explained based on these m principal components.

$$\frac{\sum_{k=1}^{m} \lambda_k}{\sum_{k=1}^{p} \lambda_k} = \frac{\sum_{k=1}^{m} \lambda_k}{p} 0.8,\qquad k = 1, 2, \ldots, m \ \text{ and } \ m \leq p \tag{10.9}$$

Step 7 (Construct the decision matrix $B^0 = \left[b^0_{jk}\right]_{p\times m}$). Unit vector can ensure that the importance level of indicators can be compared among different eigenvectors. Therefore, we transformed the vector b_k into a unit vector as follows.

$$\vec{b}_k^0 = \frac{\vec{b}_k}{\left|\vec{b}_k\right|} = \begin{bmatrix} b_{1k}^0 \\ b_{2k}^0 \\ \vdots \\ b_{pk}^0 \end{bmatrix}, \qquad k = 1, 2, ..., m \quad \text{and} \quad m \le p \tag{10.10}$$

where $\left|\vec{b}_k\right|$ is the magnitude (modulus) of the vector b_k. b_k^0 is the unit vector, indicating that the square sum of its components equals 1. The absolute value of the component in the vector b_k^0 reflects the importance level of the corresponding indicators. The decision matrix B^0 can be constructed based on the following equation.

$$B^0 = \left[b_{jk}^0\right]_{p\times m} = \left[\vec{b}_1^0, \vec{b}_2^0, \ldots, \vec{b}_m^0\right] = \begin{bmatrix} b_{11}^0 & b_{12}^0 & \cdots & b_{1m}^0 \\ b_{21}^0 & b_{22}^0 & \cdots & b_{2m}^0 \\ \vdots & \vdots & \ddots & \vdots \\ b_{p1}^0 & b_{p2}^0 & \cdots & b_{pm}^0 \end{bmatrix} \tag{10.11}$$

To be noticed, the results obtained using SPSS software are the 'component matrix' ($F = \left[\vec{f}_1, \vec{f}_2, \ldots, \vec{f}_m\right]_{p\times m}$) with the loadings as its elements instead of the decision matrix ($B^0 = \left[\vec{b}_1^0, \vec{b}_2^0, \ldots, \vec{b}_m^0\right]_{p\times m}$). The relationship between the 'component matrix' and the decision matrix is as follows.

$$\vec{b}_k = \frac{\vec{f}_k}{\sqrt{\lambda_k}}, \qquad k = 1, 2, ..., m \quad \text{and} \quad m \le p \tag{10.12}$$

Step 8 (Construct the primary component models). The principal component is a linear combination of all the indicators that go through the origin. The m principal components of DMU i can be formulated as follows.

$$\begin{cases} F_{i1} = b_{11}^0 z_{i1} + b_{21}^0 z_{i2} + \cdots + b_{p1}^0 z_{ip} \\ F_{i2} = b_{12}^0 z_{i1} + b_{22}^0 z_{i2} + \cdots + b_{p2}^0 z_{ip} \\ \quad\vdots \\ F_{im} = b_{pm}^0 z_{i1} + b_{p2}^0 z_{i2} + \cdots + b_{pm}^0 z_{ip} \end{cases} \tag{10.13}$$

where $F_{i1}, F_{i2}, \ldots, F_{im}$ are the m principal components of DMU i. $z_{i1}, z_{i2}, \ldots, z_{ip}$ indicates the value of indicators from 1 to p for DMU i. The rank of principal components ($F_{i1}, F_{i2}, \ldots, F_{im}$) is based on the extent of variance included in the components measured by eigenvalue ($\lambda_1 \ge \lambda_1 \ge \cdots \ge \lambda_m$). In other words, F_{i1} includes the maximum variance of the dataset, while F_{im} has the minimum variance. These principal components are not correlated but perpendicular to each other, indicating that they contain the information of different 'statistical dimensions'.

Step 9 (Evaluate the performance of DMUs). The study constructs the comprehensive component model of DMU i as follows (Zhu 1998).

$$F_i = \sum_{k=1}^{m} W_k \times F_{ik} = a_1 z_{i1} + a_2 z_{i2} + \cdots + a_p z_{ip},$$
$$i = 1, 2, \ldots, n, \quad k = 1, 2, \ldots, m \quad \text{and} \quad m \le p \tag{10.14}$$

$$W_k = \begin{cases} \lambda_k / \sum_{k=1}^{m} \lambda_k, & if \quad \sum_{j=1}^{p} b_{jk}^0 \ge 0 \\ -\lambda_k / \sum_{k=1}^{m} \lambda_k, & if \quad \sum_{j=1}^{p} b_{jk}^0 < 0 \end{cases} \tag{10.15}$$

The percentage of variance explained by each component represents its relative importance. Therefore, the study uses the $\lambda_k / \sum_{k=1}^{m} \lambda_k$ as the weight of the primary component k, which is denoted as W_k. The sign of W_k depends on the $\sum_{j=1}^{p} b_{jk}^0$, and a negative sign will be added to W_k once $\sum_{j=1}^{p} b_{jk}^0$ is negative. The sign change does not change the explanation of the principal component (Zhu 1998). a_j means the importance level of the indicator j. F_i indicates the performance scores of DMU i. The higher the F_i, the better the performance is. z_{ij} is the standardized value of the indicator j for DMU i based on step 3.

Step 10 (Standardize the performance of DMUs to make scores range from 0 to 100). The performance score of DMUs based on step 9 can be negative or positive, which makes it not straightforward to compare and demonstrate. Therefore, we further standardize the scores to make them range from 0 to 100. The higher the standardized scores of a DMU ($F_i^{'}$), the better the sustainability is. 0 indicates the worst performance, while 100 means the best performance. The standardization process is as follows.

$$F_i^{'} = 100 \times \frac{F_i - \min\limits_{1 \le i \le n}(F_i)}{\max\limits_{1 \le i \le n}(F_i) - \min\limits_{1 \le i \le n}(F_i)} \tag{10.16}$$

where $\min\limits_{1 \le i \le n}(F_i)$ and $\max\limits_{1 \le i \le n}(F_i)$ indicate the minimum and maximum value among all the F_i of DMU i.

10.2.2 Sample Cities and Data

Table 10.1 presents the indicators of three aspects, including economy, environment, and society. 14 indicators are included in this study. The data of these indicators are collected from the statistical yearbook of each city. The innovation capacity indicates

Table 10.1 List of indicators regarding economy, society, and environment

Aspect	No	Indicator	Unit	Dimension
Economy	1	GDP per capita	10,000 Yuan /person	Positive
	2	GDP growth rate	%	Positive
	3	Share of tertiary industry	%	Positive
Society	4	Share of the unemployed person	%	Negative
	5	Ratio of teacher to total population	%	Positive
	6	Share of foreign direct investments to GDP	%	Positive
	7	Innovation capacity	Unit/person	Positive
	8	Student–teacher ratio of regular higher education institutions	–	Positive
	9	Number of public transportation vehicles per 10,000 persons	Unit /10^4 persons	Positive
Environment	10	CO_2 emissions per capita	Tonnes /person	Negative
	11	CO_2 emissions intensity	Tonnes/10^4 ¥	Negative
	12	Volume of industry sulfur dioxide per capita	Tonne /person	Negative
	13	Ratio of industrial solid wastes comprehensively utilized	%	Positive
	14	Ratio of waste water centralized treated of sewage work	%	Positive

the patent per capita. The patent data are collected from the State Intellectual Property Office of China, which includes invention patents, utility model patents, and design patents. The GDP has been converted to the 2005 constant price based on the GDP index. CO_2 emissions are sourced from Chen et al. (2020).

According to 'The Plan for the Sustainable Development of Chinese Resource-based Cities (2013–2020)' issued by the State Council, there are 262 resource-based cities in China, among which 69 are resource-exhausted. In this study, owing to the data unavailability at the county level, we only take the prefectural level cities into consideration. A prefectural-level city doesn't indicate the usual meaning of the term (i.e., urban settlement), but means an administrative unit that with not only urban area but rural area. Daxinganling city is also not included because of insufficient data. Thus, a total of 24 prefectural level cities are studied and some descriptions of these 24 cities can be found in Table 10.2.

Table 10.2 List of resource-exhausted cities at the prefectural level of China

No	Province	City	Note	Establishment time
1	Inner Mongolia	Wuhai	Coal	2011
2	Liaoning	Fushun	Coal	2009
3	Liaoning	Fuxin	Coal	2008
4	Liaoning	Panjin	Crude oil	2008
5	Jilin	Liaoyuan	Coal	2008
6	Jilin	Baishan	Coal	2008
7	Heilongjiang	Hegang	Coal	2011
8	Heilongjiang	Shuangyashan	Coal	2011
9	Heilongjiang	Yichun	Forest	2008
10	Heilongjiang	Qitaihe	Coal	2009
11	Anhui	Huaibei	Coal	2009
12	Anhui	Tongling	Copper	2009
13	Jiangxi	Jingdezhen	Porcelain	2009
14	Jiangxi	Pingxiang	Coal	2008
15	Jiangxi	Xinyu	Iron	2011
16	Shandong	Zaozhuang	Coal	2009
17	Henan	Jiaozuo	Coal	2008
18	Henan	Puyang	Crude oil	2011
19	Hubei	Huangshi	Iron, copper, coal, and wollastonite	2009
20	Guangdong	Shaoguan	Coal and iron	2011
21	Sichuan	Luzhou	Natural gas	2011
22	Tibet	Tongchuan	Coal	2009
23	Gansu	Baiyin	Silver and copper	2008
24	Ningxia	Shizuishan	Coal	2008

10.3 Structure Detection of Dataset

Table 10.3 shows the results of Bartlett's test and Kaiser–Meyer–Olkin measure, which are tested using the SPSS software. The Kaiser–Meyer–Olkin measure is a

Table 10.3 Results of the Kaiser–Meyer–Olkin measure

KMO and Bartlett's Test		
Kaiser–Meyer–Olkin Measure of Sampling Adequacy		0.630
Bartlett's Test of Sphericity	Approx. Chi-Square	1620.822
	df	91
	Sig	0.000

Table 10.4 Interpretation of the results using Kaiser–Meyer–Olkin measure (Dziuban and Shirkey 1974)

KMO measure	Meaning
KMO ≥ 0.9	Marvellous
0.8 ≤ KMO < 0.9	Meritorious
0.7 ≤ KMO < 0.8	Middling
0.6 ≤ KMO < 0.7	Mediocre
0.5 ≤ KMO < 0.6	Miserable
KMO < 0.5	Unacceptable

test that shows the share of variance in the indicators which could be caused by underlying factors (Dziuban and Shirkey 1974). The result of the Kaiser–Meyer–Olkin measure ranges from 0 to 1 (Dziuban and Shirkey 1974). It is generally more convincing to conduct PCA analysis if the value of the result becomes higher. If the value is larger than 0.6, indicating the sample complies with the requirement of data structure and can use the PCA (Dziuban and Shirkey 1974). For Bartlett's test of sphericity, the null hypothesis is that the correlation matrix of the dataset is an identity matrix. If it cannot reject the null hypothesis, it means that your indicators are not related and thus it is not appropriate for structure detection. Small values (less than 0.05) of the significance level indicate that factor analysis may be useful with this dataset.

The results in Table 10.3 show that it is suitable for the dataset to conduct structure detection since the value of the Kaiser–Meyer–Olkin measure is 0.630 and the null hypothesis is rejected at a 1% significant level.

Table 10.4 can be a reference to show the meaning of value based on the Kaiser–Meyer–Olkin measure.

Table 10.5 shows the correlation matrix of 14 indicators. The order of the variables shown in Table 10.5 is the same as in Table 10.1. The bottom left triangle indicates the Pearson correlation, while the top right triangle means the Spearman correlation. The eigenvalue and eigenvector are based on the correlation matrix measured by the Pearson correlation and this matrix is symmetric. The symbols *, **, and *** indicates the value is significant at 10%, 5%, and 1% levels, respectively. The value in the matrix shows that many indicators are significantly correlated with each other, and the PCA can be used to reduce the correlation and simplify the dimensions.

10.4 The Weights and Sustainability Performance Based on the Principal Component Analysis

Table 10.6 shows the communalities of 14 indicators. Extraction commonalities can be used to evaluate the variance of each indicator that can be extracted by the factors in the PCA. If the extraction is low, this indicator is not fit well with the factor solution and can be discarded in further analysis. The results in Table 10.6 show that the initial

Table 10.5 Correlation matrix of the 14 variables

	var1	var2	var3	var4	var5	var6	var7	var8	var9	var10	var11	var12	var13	var14
var1	1	−0.499 ***	0.05	−0.249 ***	−0.195 ***	0.196 ***	0.538 ***	−0.137 **	0.483 ***	−0.559 ***	0.182 ***	−0.003	0.135 **	0.579 ***
var2	−0.370 ***	1	−0.278 ***	0.058	0.239 ***	0.027	−0.456 ***	0.034	−0.118 **	0.295 ***	−0.059	−0.217 ***	−0.237 ***	-0.428 ***
var3	0.104 *	−0.276 ***	1	−0.368 ***	−0.103 *	−0.094	0.057	−0.017	0.115 *	−0.098 *	−0.078	0.008	−0.028	−0.008
var4	-0.249 ***	0.076	−0.271 ***	1	0.094	−0.126 **	0.094	−0.079	−0.463 ***	0.462 ***	0.369 ***	0.215 ***	0.242 ***	−0.022
var5	−0.110 *	0.202 ***	0.015	0.033	1	−0.149 **	−0.222 ***	0.253 ***	−0.211 ***	0.08	−0.330 ***	−0.284 ***	−0.218 ***	−0.009
var6	0.188 ***	0.063	−0.133 **	−0.051	−0.105 *	1	0.247 ***	−0.126 **	0.049	0.185 ***	0.397 ***	0.053	0.201 ***	0.183 ***
var7	0.408 ***	−0.300 ***	0.145 **	0.082	−0.093	0.129 **	1	−0.348 ***	−0.119 **	−0.117 **	0.385 ***	0.297 ***	0.298 ***	0.438 ***
var8	−0.102 *	0.046	0.006	−0.056	0.185 ***	−0.104 *	−0.244 ***	1	−0.018	−0.186 ***	−0.345 ***	0.088	−0.075	−0.099 *
var9	0.577 ***	−0.133 **	0.123 **	−0.453 ***	−0.138 **	−0.024	−0.039	−0.033	1	−0.440 ***	−0.140 **	−0.290 ***	−0.009	0.143 **
var10	−0.711 ***	0.281 ***	−0.100 *	0.382 ***	−0.051	0.088	0.01	−0.053	−0.626 ***	1	0.674 ***	0.185 ***	0.198 ***	−0.302 ***
var11	0.055	0.038	−0.077	0.317 ***	−0.300 ***	0.311 ***	0.308 ***	−0.297 ***	−0.216 ***	0.539 ***	1	0.282 ***	0.382 ***	0.101 *

(continued)

Table 10.5 (continued)

	var1	var2	var3	var4	var5	var6	var7	var8	var9	var10	var11	var12	var13	var14
var12	−0.269 ***	−0.074	0.042	0.342 ***	−0.227 ***	0.159 ***	0.217 ***	−0.046	−0.487 ***	0.589 ***	0.515 ***	1	0.322 ***	0.03
var13	0.043	-0.204 ***	−0.039	0.191 ***	−0.302 ***	0.120 **	0.123 **	−0.328 ***	−0.095	0.192 ***	0.362 ***	0.339 ***	1	0.326 ***
var14	0.496 ***	−0.316 ***	0.007	−0.056	−0.04	0.180 ***	0.346 ***	−0.105 *	0.116 **	−0.249 ***	0.119 **	−0.041	0.254 ***	1

Table 10.6 Communalities of 14 indicators

No	Indicator	Initial	Extraction
1	GDP per capita	1	0.887
2	GDP growth rate	1	0.736
3	Share of tertiary industry	1	0.842
4	Share of unemployed person	1	0.760
5	Ratio of teacher to total population	1	0.854
6	Share of foreign direct investments to GDP	1	0.830
7	Innovation capacity	1	0.817
8	Student–teacher ratio of regular higher education institutions	1	0.896
9	Number of public transportation vehicles per 10,000 persons	1	0.767
10	CO_2 emissions per capita	1	0.867
11	CO_2 emissions intensity	1	0.764
12	Volume of industry sulfur dioxide per capita	1	0.766
13	Ratio of industrial solid wastes comprehensively utilized	1	0.822
14	Ratio of wastewater centralized treated of sewage work	1	0.750

variance included in the dataset is 1 for all indicators. The extraction commonalities in the PCA are fit well, with the lowest level at 0.736 for the indicator 'GDP growth rate'.

Table 10.7 shows the total variance explained for each component. The leftmost section of Table 10.7 demonstrates the variance explained by the 14 components. There are 14 components in total, among which five components have eigenvalues larger than 1. The share of variance is equal to the share of the corresponding eigenvalue in the total eigenvalues considered. The total value of the eigenvalues is 14, and thus the share of variance for the first component is 24.226%, which means that 24.226% of the variance in the dataset can be explained by this component. For the second component, 19.541% of the variance can be explained using this component. The share of cumulative variance explained by the first and the second components is 43.767%.

The second section of Table 10.7 shows the variance that can be explained by the extracted components before conducting rotation. The cumulative variability explained by the top seven components in the extracted solution is about 81.131%, which is the same as the initial solution. Thus, there is no loss in the variation explained by the initial solution. In the case there is a difference between the values in the first section and the second section, it can be interpreted that the latent components are unique to the original variables and variability that simply cannot be explained by the component model. This study considers the top seven components as the principal component, which take up 81.131% of the variability of the dataset. The rightmost section of this table shows the variance explained by the extracted factors after rotation. The rotated factor model makes some small adjustments for all these 7 components.

Table 10.7 Total variance explained

	Initial eigenvalues			Extraction sums of squared loadings			Rotation sums of squared loadings		
Comp.	Total	% of Variance	Cumulative %	Total	% of Variance	Cumulative %	Total	% of Variance	Cumulative %
1	3.392	24.226	24.226	3.392	24.226	24.226	3.122	22.298	22.298
2	2.736	19.541	43.767	2.736	19.541	43.767	1.586	11.329	33.626
3	1.326	9.471	53.238	1.326	9.471	53.238	1.491	10.65	44.276
4	1.182	8.439	61.677	1.182	8.439	61.677	1.36	9.712	53.988
5	1.032	7.375	69.052	1.032	7.375	69.052	1.294	9.246	63.234
6	0.918	6.558	75.609	0.918	6.558	75.609	1.26	9.002	72.237
7	0.773	5.522	81.131	0.773	5.522	81.131	1.245	8.895	81.131
8	0.587	4.19	85.322						
9	0.552	3.943	89.265						
10	0.438	3.128	92.393						
11	0.401	2.865	95.258						
12	0.338	2.411	97.669						
13	0.269	1.921	99.59						
14	0.057	0.41	100						

The results of the decision matrix can be found in Table 10.8. There are seven principal components considered, each of which has an eigenvalue and an eigenvector (Table 10.8). Table 10.9 shows the component matrix obtained using SPSS software instead of the MATLAB code shown in Sect. 10.6. The results obtained using SPSS software are the 'component matrix' ($F = \left[f_1, f_2, \dots f_m\right]_{p\times m}$) instead of the decision matrix ($B^0 = \left[b_1^0, b_2^0, \dots b_p^0\right]_{p\times m}$). The 'component matrix' can be transferred to the decision matrix based on $\vec{b}_k = \frac{\vec{f}_k}{\sqrt{\lambda_k}}$, as shown in step 7 in Sect. 10.2.1. Based on Table 10.8, the first principal component (PC1) is mainly explained by Var 10 (0.490), followed by Var 9 (–0.410), and Var 12 (0.394). Var 10, Var 9, and Var 12 indicate 'CO_2 emissions per capita', Number of public transportation vehicles per 10,000 persons', and 'Volume of industry sulfur per capita', respectively. For the second principal component (PC2), it is primarily composed of Var 7 (–0.394), Var 14 (–0.369), and Var 1 (–0.369). Var 7, Var 14, and Var 1 means 'Innovation capacity', 'Ratio of wastewater centralized treated of sewage work', and ' GDP per capita', respectively. The top three compositions of each principal component have been highlighted in bold as shown in Table 10.8.

Table 10.10 shows the transformation matrix of the seven principal components. This matrix can be used to demonstrate the rotation level of the component matrix compared with the unrotated component matrix. If the values of off-diagonal elements are small, it indicates there are smaller rotations between these two matrices. By contrast, the large values in the off-diagonal elements mean larger rotations.

Table 10.11 shows the sustainability performance of 24 resource-exhausted cities from 2005 to 2016, while Table 10.12 shows the sustainability performance after

Table 10.8 Decision matrix of the seven principal components

	PC1	PC2	PC3	PC4	PC5	PC6	PC7
Eigenvalue	3.392	2.736	1.326	1.182	1.032	0.918	0.773
Var1	−0.355	**−0.369**	0.127	−0.160	0.040	0.028	0.210
Var2	0.149	0.313	**0.463**	0.023	0.191	−0.276	0.034
Var3	−0.122	−0.075	**−0.618**	0.176	**0.383**	−0.246	−0.183
Var4	0.327	−0.022	0.020	−0.324	**−0.437**	0.023	0.308
Var5	−0.061	0.276	−0.044	**−0.599**	0.066	**−0.337**	−0.356
Var6	0.072	−0.218	**0.453**	−0.086	**0.555**	0.190	−0.258
Var7	0.014	**−0.394**	−0.178	**−0.346**	0.154	**−0.292**	**0.368**
Var8	−0.070	0.276	−0.175	−0.244	0.220	**0.733**	0.147
Var9	**−0.410**	−0.095	0.179	0.296	0.016	−0.024	0.182
Var10	**0.491**	0.057	−0.060	0.055	0.166	−0.060	−0.048
Var11	0.322	−0.330	0.155	0.055	0.188	−0.082	0.218
Var12	**0.394**	−0.184	−0.234	0.026	0.174	0.211	0.041
Var13	0.177	−0.337	0.015	0.225	−0.367	0.102	**−0.504**
Var14	−0.133	**−0.369**	0.058	**−0.384**	−0.107	0.155	**−0.370**

Table 10.9 Component Matrix

Variable	Component 1	2	3	4	5	6	7
GDP per capita	−0.653	0.61	0.147	0.174	0.041	0.027	−0.185
GDP growth rate	0.274	−0.518	0.533	−0.025	0.194	−0.265	−0.03
Share of tertiary industry	−0.225	0.124	−0.712	−0.191	0.389	−0.235	0.161
Share of unemployed person	0.603	0.036	0.023	0.352	−0.444	0.022	−0.271
Ratio of teacher to total population	−0.113	−0.457	−0.051	0.651	0.067	−0.323	0.313
Share of foreign direct investments to GDP	0.132	0.36	0.521	0.093	0.564	0.182	0.227
Innovation capacity	0.025	0.652	−0.205	0.377	0.156	−0.28	−0.324
Student–teacher ratio of regular higher education institutions	−0.129	−0.456	−0.201	0.266	0.223	0.702	−0.129
Number of public transportation vehicles per 10,000 persons	−0.755	0.158	0.206	−0.322	0.016	−0.023	−0.16
CO_2 emissions per capita	0.904	−0.094	−0.069	−0.06	0.169	−0.058	0.042
CO_2 emissions intensity	0.592	0.546	0.178	−0.06	0.191	−0.079	−0.192
Volume of industry sulfur dioxide per capita	0.726	0.304	−0.27	−0.029	0.176	0.202	−0.036
Ratio of industrial solid wastes comprehensively utilized	0.326	0.557	0.017	−0.245	−0.373	0.098	0.443
Ratio of waste water centralized treated of sewage work	−0.244	0.61	0.067	0.417	−0.108	0.148	0.325

standardization. The average sustainability of these cities is 53.783 in 2015, while the level increases to 64.297 in 2016 (see Table 10.12). However, there is still much room for improvement. Among the 24 cities, Shuangyashan, Fushun, Fuxin, Shizuishan, and Wuhai see the most significant risk in sustainability, at a mere 52.242, 52.447, 47.371, 34.062, and 4.113 in 2016 (see Table 10.12).

Figure 10.1 compares the sustainability level of the 24 resource-exhausted cities in 2005 with that in 2016. Most of the resource-exhausted cities see an increase in sustainability from 2005 to 2016, except Panjin, Shuangyashan, and Yichun-HLJ (see Fig. 10.1). This indicates that Panjin, Shuangyashan, and Yichun-HLJ face significant sustainability risks and prompt actions should be taken to reverse this deterioration. Panjin, Shuangyashan, and Yichun-HLJ are located in Liaoning province, Heilongjiang province, and Heilongjiang province, respectively. These two provinces belong to Northeast China and are used to be the old industrial base of China. Because

Table 10.10 Component transformation matrix

Component	1	2	3	4	5	6	7
1	0.935	−0.122	−0.006	0.136	0.241	0.087	0.164
2	−0.047	0.583	0.549	0.397	−0.065	0.28	0.341
3	−0.297	−0.217	−0.173	0.009	0.657	0.586	0.242
4	0.105	0.244	0.466	−0.71	0.32	0.066	−0.319
5	0.082	−0.374	0.197	−0.045	−0.536	0.673	−0.269
6	0.054	0.373	−0.275	0.396	0.197	0.166	−0.748
7	0.121	0.51	−0.581	−0.401	−0.279	0.291	0.255

Extraction Method: Principal Component Analysis

Rotation Method: Varimax with Kaiser Normalization

of the decline of its once-powerful resource-related sectors, the Northeast region is named the Rust Belt of China (Campbell 2005; Xiao et al. 2019a), whose development path is different from the Yangtze River Delta region (Xiao et al. 2019b). Many cities in Northeast China used to be abundant in natural resources. However, after decades of the exploitation of natural resources, many cities are facing resource depletion problems and Panjin, Shuangyashan, and Yichun-HLJ are the typical cases. The sustainability risks of these cities threaten the long-term development and integrated measures should be taken to improve sustainability and reduce the risk of deterioration.

10.5 Conclusions and Policies for Mitigating Sustainability Risks

Considering the significant sustainability risks faced by the resource-exhausted cities, the study constructs an evaluation framework covering economic, social, and environmental dimensions to evaluate the sustainability of resource-exhausted cities and identify potential risks. The PCA is used to evaluate the sustainability risks of 24 resource-exhausted cities in China from 2005 to 2016. The main findings are as follows:

The sustainability of resource-exhausted cities sees an increasing trend. The average sustainability of these cities is 53.783 in 2015, while the level increases to 64.297 in 2016. However, there is still much room for improvement. Among the 24 cities, Shuangyashan, Fushun, Fuxin, Shizuishan, and Wuhai see the most significant risk in sustainability, at a mere 52.242, 52.447, 47.371, 34.062, and 4.113 in 2016.

Panjin, Shuangyashan, and Yichun-HLJ are the only three cities whose sustainability sees a decreasing trend from 2005 to 2016. This indicates that Panjin, Shuangyashan, and Yichun-HLJ face relatively significant sustainability risks and prompt actions should be taken to reverse this deterioration. These three cities are

Table 10.11 The sustainability scores of 24 resource-exhausted cities from 2005 to 2016

No	City	2005	2006	2007	2008	2009	2010	2011	2012	2013	2014	2015	2016
1	Wuhai	−2.799	−2.168	−2.054	−1.879	−1.707	−1.361	−1.632	−1.604	−1.906	−2.335	−2.860	−2.673
2	Fushun	−0.833	−0.646	−0.234	−0.465	−0.124	0.020	−0.336	−0.510	−0.034	−0.092	−0.344	−0.525
3	Fuxin	−0.744	−0.521	−0.469	−0.280	−0.183	0.108	0.034	−0.129	−0.196	0.027	−0.411	−0.710
4	Panjin	0.005	−0.082	0.031	−0.011	0.294	0.858	1.678	1.163	0.899	0.253	−0.088	−0.279
5	Liaoyuan	−0.613	−0.576	−0.263	−0.220	0.277	0.620	0.352	0.360	0.486	0.348	0.342	0.518
6	Baishan	−0.775	−0.610	−0.290	−0.408	−0.417	−0.307	−0.075	−0.351	−0.099	−0.138	−0.030	0.260
7	Hegang	−1.058	−1.050	−0.895	−0.856	−0.352	−0.168	−0.415	−0.367	−0.512	−0.498	−0.381	−0.478
8	Shuangyashan	−0.188	−0.172	0.051	−0.046	−0.024	−0.024	0.089	0.047	−0.073	−0.412	−0.354	−0.489
9	Yichun-HLJ	−0.428	−0.396	−0.245	−0.189	0.073	0.276	0.258	0.078	−0.065	−0.324	−0.459	−0.445
10	Qitaihe	−0.309	−0.320	−0.342	−0.276	−0.202	−0.109	0.157	0.100	−0.332	−0.419	−0.469	−0.191
11	Huaibei	−0.296	−0.318	−0.073	0.122	0.259	0.475	0.653	0.776	0.890	0.916	0.982	1.188
12	Tongling	0.272	0.543	0.315	0.715	0.890	0.870	0.687	0.831	0.898	0.764	0.862	0.982
13	Jingdezhen	0.309	−0.120	0.251	0.052	0.521	0.655	0.499	0.479	0.455	0.439	0.453	0.577
14	Pingxiang	0.039	−0.006	0.423	0.656	0.671	0.722	0.595	0.606	0.584	0.590	0.599	0.797
15	Xinyu	0.031	0.187	0.495	0.937	0.725	0.928	0.890	0.736	0.341	0.409	0.393	0.444
16	Zaozhuang	−0.032	0.381	0.258	0.594	0.679	0.692	0.566	0.556	0.490	0.475	0.419	0.399
17	Jiaozuo	0.075	0.123	0.367	0.165	0.482	0.642	0.754	0.778	0.745	0.699	0.826	1.010
18	Puyang	0.457	0.497	0.096	0.497	0.552	0.734	0.815	0.995	1.035	0.985	1.047	1.061
19	Huangshi	−0.192	0.008	−0.020	0.099	0.227	0.176	0.395	0.553	0.695	0.732	0.268	0.355

(continued)

Table 10.11 (continued)

No	City	2005	2006	2007	2008	2009	2010	2011	2012	2013	2014	2015	2016
20	Shaoguan	0.091	0.298	0.436	0.426	0.233	0.364	0.464	0.182	0.238	0.316	0.204	0.153
21	Luzhou	0.444	0.442	0.601	0.181	0.161	0.373	0.449	0.516	0.586	0.642	0.731	0.833
22	Tongchuan	−0.617	−0.399	−0.552	−0.489	−0.101	0.084	0.222	0.306	0.232	0.114	0.045	0.065
23	Baiyin	−1.246	−1.081	−0.856	−0.834	−0.810	−0.613	−0.513	−0.432	−0.202	−0.267	−0.278	−0.144
24	Shizuishan	−1.650	−1.635	−1.248	−1.168	−1.630	−1.335	−0.879	−0.812	−0.607	−0.682	−1.142	−1.314
	Average	−0.419	−0.318	−0.176	−0.112	0.021	0.195	0.238	0.202	0.189	0.106	0.015	0.058

Table 10.12 The standardized scores of sustainability of 24 resource-exhausted cities from 2005 to 2016

No	City	2005	2006	2007	2008	2009	2010	2011	2012	2013	2014	2015	2016
1	Wuhai	1.339	15.235	17.753	21.608	25.391	33.033	27.059	27.676	21.005	11.557	0.000	4.113
2	Fushun	44.666	48.789	57.861	52.760	60.283	63.469	55.623	51.775	62.273	60.999	55.446	51.448
3	Fuxin	46.629	51.528	52.687	56.855	58.981	65.390	63.758	60.180	58.693	63.621	53.962	47.371
4	Panjin	63.128	61.200	63.699	62.769	69.489	81.935	100.000	88.643	82.830	68.599	61.086	56.862
5	Liaoyuan	49.512	50.335	57.225	58.166	69.122	76.672	70.777	70.961	73.730	70.676	70.551	74.432
6	Baishan	45.941	49.581	56.617	54.037	53.823	56.258	61.357	55.284	60.827	59.977	62.351	68.742
7	Hegang	39.712	39.888	43.296	44.157	55.258	59.305	53.864	54.929	51.734	52.051	54.614	52.483
8	Shuangyashan	58.876	59.221	64.135	62.008	62.491	62.493	64.986	64.048	61.409	53.933	55.223	52.242
9	Yichun-HLJ	53.587	54.289	57.622	58.851	64.636	69.110	68.702	64.731	61.584	55.868	52.907	53.221
10	Qitaihe	56.198	55.963	55.491	56.934	58.567	60.610	66.468	65.219	55.702	53.791	52.675	58.813
11	Huaibei	56.502	56.012	61.418	65.702	68.731	73.485	77.417	80.117	82.627	83.207	84.663	89.200
12	Tongling	69.011	74.993	69.955	78.771	82.630	82.190	78.156	81.326	82.804	79.860	82.008	84.654
13	Jingdezhen	69.827	60.362	68.554	64.156	74.503	77.453	74.004	73.583	73.054	72.681	73.000	75.723
14	Pingxiang	63.877	62.875	72.331	77.476	77.793	78.917	76.135	76.376	75.877	76.018	76.209	80.572
15	Xinyu	63.691	67.149	73.934	83.661	79.000	83.459	82.630	79.234	70.529	72.035	71.670	72.801
16	Zaozhuang	62.305	71.418	68.703	76.100	77.989	78.268	75.501	75.260	73.816	73.483	72.247	71.806
17	Jiaozuo	64.665	65.721	71.107	66.652	73.634	77.156	79.638	80.161	79.433	78.422	81.224	85.284
18	Puyang	73.085	73.964	65.135	73.965	75.178	79.189	80.984	84.941	85.834	84.722	86.090	86.400
19	Huangshi	58.782	63.204	62.575	65.190	68.015	66.890	71.721	75.197	78.329	79.148	68.913	70.844
20	Shaoguan	65.024	69.589	72.635	72.411	68.155	71.044	73.239	67.023	68.261	69.975	67.519	66.383
21	Luzhou	72.801	72.752	76.260	66.999	66.567	71.238	72.908	74.397	75.941	77.167	79.126	81.383

(continued)

Table 10.12 (continued)

No	City	2005	2006	2007	2008	2009	2010	2011	2012	2013	2014	2015	2016
22	Tongchuan	49.426	54.221	50.863	52.237	60.797	64.876	67.909	69.750	68.132	65.520	63.998	64.442
23	Baiyin	35.560	39.196	44.163	44.637	45.159	49.514	51.702	53.489	58.558	57.126	56.897	59.852
24	Shizuishan	26.656	26.990	35.515	37.269	27.102	33.601	43.654	45.116	49.650	47.992	37.841	34.063
	Average	53.783	56.020	59.147	60.557	63.471	67.315	68.258	67.476	67.193	65.351	63.343	64.297

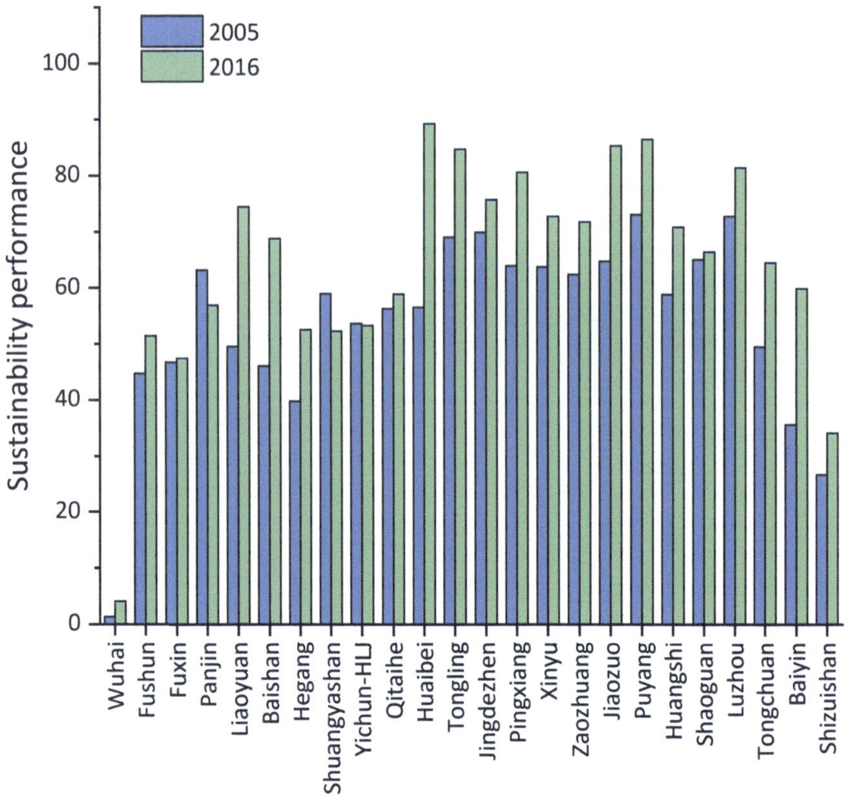

Fig. 10.1 The sustainability scores of 24 resource-exhausted cities in 2005 and 2016. The value in the figure is based on Table 10.12

located in the Northeast region, the Rust Belt of China. Many cities in Northeast China used to be abundant in natural resources. However, after decades of the exploitation of natural resources, many cities are facing resource depletion problems, and Panjin, Shuangyashan, and Yichun-HLJ are the typical cases. The sustainability risks of these cities could threaten the long-term development and integrated measures needed to be taken to improve sustainability and reduce the risk of deterioration.

Some policies are proposed to mitigate the sustainability risks of resource-exhausted cities. First, policy-makers should keep a close eye on the economic, social, and environmental development of resource-exhausted cities. The evaluation framework used in this study can be a reference to take stock of where resource-exhausted cities stand in terms of sustainability, identify the potential risks, and further promote sustainable development. Second, for the laggard cities in sustainability, relevant policies should be implemented to mitigate the sustainability risks and then reverse the deterioration process. For the other cities, they should plan in advance to avoid getting into similar situations.

10.6 Code Availability

The MATLAB code used to evaluate the performance of the DUMs using principal component analysis (PCA) is published for transparency, as follows.

```
% Evaluating and ranking the performance of DMU using the Principal Component Analysis
% Step 1: Import data and construct the origin matrix Y
% Note (1): Please put the 'Table.xlsx' file in the same path as the code file.
% Note (2): Please make sure all the indicators are in positive dimensions
% Note (3): The format of excel file: the first row is variable names.
clear
Y=xlsread('Table.xlsx','Sheet1');

% Step 2: Standardize the matrix Y and obtain matrix Z
Z=zscore(Y)

% Step 3: Construct the covariance matrix R
R = cov(Z);

% Step 4: Calculate the eigenvalue (lamda) and eigenvector (B)
% Note (1): The eigenvector has been automatically transformed to unit vector
% Note (2): The code 'svd' and 'eig' return results in different order (one sorted large to small, the other in
reverse):
[B,lamda_matrix] = svd(R);
lamda = diag(lamda_matrix);

% Step 5: Determine the number of principal components to be considered
fprintf('Cumulative variance: %6.3f \n', cumsum(lamda)/sum(lamda));
k = find((cumsum(lamda)/sum(lamda)) > 0.8, 1, 'first');

% Step 6: Construct the decision matrix b
b = B(:, 1:k);

% Step 7: Construct the comprehensive component models
W = zeros(k, 1);
Sum_b = sum(b);
for i=1:k
    if Sum_b(i)>0
        W(i)=lamda(i) / sum(lamda(1:k));
    else
        W(i)=-lamda(i) / sum(lamda(1:k));
    end
end
a = zeros(size(Y,2), 1);
for i=1:size(Y,2)
    a(i) = a(i) + b(i,:)*W;
end
```

```
% Step 8: Evaluate the performance score of the DMU
score = Z * a
% Step 9: Standardize the performance of DMUs to make scores range from 0 to 100
for i=1:size(Y,1)
    score2(i) = 100*(score(i)-min(score))/(max(score)-min(score));
end
fprintf('Scores is: %6.3f \n', score2);
```

Acknowledgements Huijuan XIAO would like to express her sincere thanks to the Research Committee of The Hong Kong Polytechnic University for the financial support of the project through a Ph.D. studentship (project account code: RK2K).

References

Auty RM, Auty RM, Mikesell RF (1998) Sustainable Development in Mineral Economies. Clarendon

Brundtland GH, Khalid M, Agnelli S, Al-Athel S, Chidzero B (1987) Our common future, p 8. New York

Campbell C (2005) Swallows and Settlers: The Great Migration from North China to Manchuria. J Asian Stud 64:1004–1005. https://doi.org/10.1017/S0021911805002469

Chen J, Gao M, Cheng S, Hou W, Song M, Liu X, Liu Y, Shan Y (2020) County-level CO 2 emissions and sequestration in China during 1997–2017. Scientific Data 7:391. https://doi.org/10.1038/s41597-020-00736-3

Dong S, Li Z, Li B, Xue M (2007) Problems and Strategies of Industrial Transformation of China's Resource-based Cities. China Popul Resour Environ 17:12–17. https://doi.org/10.1016/S1872-583X(08)60005-4

Dziuban CD, Shirkey EC (1974) When is a correlation matrix appropriate for factor analysis? Some decision rules. Psychol Bull 81:358–361. https://doi.org/10.1037/h0036316

He SY, Lee J, Zhou T, Wu D (2017) Shrinking cities and resource-based economy: The economic restructuring in China's mining cities. Cities 60, 75–83. https://doi.org/10.1016/j.cities.2016.07.009

Li Q, Liu S, Yang M, Xu F (2020) The effects of China's sustainable development policy for resource-based cities on local industrial transformation. Resources Policy 101940

Liou S-M, Lo S-L, Wang S-H (2004) A generalized water quality index for Taiwan. Environ Monit Assess 96:35–52

Lu C, Xue B, Lu C, Wang T, Jiang L, Zhang Z, Ren W (2016) Sustainability Investigation of resource-based cities in Northeastern China. Sustainability 8:1058. https://doi.org/10.3390/su8101058

Omrani H, Valipour M, Jafari Mamakani S (2019) Construct a composite indicator based on integrating Common Weight Data Envelopment Analysis and principal component analysis models: an application for finding development degree of provinces in Iran. Socioecon Plann Sci 68:100618. https://doi.org/10.1016/j.seps.2018.02.005

Qin B, Liu L, Tao Y (2019) Evaluation of sustainable development of resource-based cities in China. J Environ Econ 142–158. https://doi.org/10.19511/j.cnki.jee.2019.03.009

Shao S, Qi Z (2009) Energy exploitation and economic growth in Western China: An emprical analysis based on the resource curse hypothesis. Front Econ China 4:125–152. https://doi.org/10.1007/s11459-009-0008-1

Shao S, Zhang Y, Tian Z, Li D, Yang L (2020) The regional Dutch disease effect within China: a spatial econometric investigation. Energy Economics 88:104766. https://doi.org/10.1016/j.eneco.2020.104766

Szalai L (2018) A review on the resource curse. Period Polytech Soc Manag Sci 26:179–189. https://doi.org/10.3311/PPso.10021

The Chinese Government (2011) The Resource-exhausted cities list in China

The Chinese Government (2008) Suggestions on promoting the development of resource-exhausted cities in China

Vega M, Pardo R, Barrado E, Debán L (1998) Assessment of seasonal and polluting effects on the quality of river water by exploratory data analysis. Water Res 32:3581–3592

Xiao H, Duan Z, Zhou Y, Zhang N, Shan Y, Lin X, Liu G (2019a) CO2 emission patterns in shrinking and growing cities: A case study of Northeast China and the Yangtze River Delta. Appl Energy 251:113384. https://doi.org/10.1016/j.apenergy.2019.113384

Xiao H, Shan Y, Zhang N, Zhou Y, Wang D, Duan Z (2019b) Comparisons of CO2 emission performance between secondary and service industries in Yangtze River Delta cities. J Environ Manage 252:109667. https://doi.org/10.1016/j.jenvman.2019.109667

Zhang H, Xiong L, Li L, Zhang S (2018) Political incentives, transformation efficiency and resource-exhausted cities. J Clean Prod 196:1418–1428. https://doi.org/10.1016/j.jclepro.2018.06.093

Zhang Q, Brouwer R (2020) Is China affected by the resource curse? A critical review of the Chinese literature. Journal of Policy Modeling 42:133–152

Zhu J (1998) Data envelopment analysis vs. principal component analysis: An illustrative study of economic performance of Chinese cities. Euro J Oper Res 111, 50–61. https://doi.org/10.1016/S0377-2217(97)00321-4s

Index

J. Ren (ed.), *Multi-Criteria Decision Analysis for Risk Assessment and Management*, Industrial Ecology and Environmental Management 1,
https://doi.org/10.1007/978-3-030-78152-1

The manufacturer's authorised representative in the EU is Springer Nature Customer Service Centre GmbH, Europaplatz 3, 69115 Heidelberg, Germany. If you have any concerns regarding our products, please contact ProductSafety@springernature.com

Printed and bound by CPI Group (UK) Ltd, Croydon, CR0 4YY
30/09/2026
02225508-0005